Computational Methods in
Genome Research

Computational Methods in Genome Research

Edited by

Sándor Suhai
German Cancer Research Centre
Heidelberg, Germany

Plenum Press ● New York and London

Library of Congress Cataloging-in-Publication Data

International Symposium on Computational Methods in Genome Research
(1992 : Heidelberg, Germany)
    Computational methods in genome research / edited by Sándor Suhai.
        p.   cm.
    "Proceedings of an International Symposium on Computational
Methods in Genome Research, held July 1-4, 1992, in Heidelberg,
Germany"--T.p. verso.
    Includes bibliographical references and index.
    ISBN 0-306-44712-6
    1. Genetics--Mathematics--Congresses. 2. Genetics--Data
processing--Congresses.   I. Suhai, Sándor. II. Title.
QH438.4.M33I58 1994
574.87'3282'0151--dc20

                                                    94-6886
                                                       CIP

Proceedings of an International Symposium on Computational Methods in Genome Research, held July 1–4, 1992, in Heidelberg, Germany

ISBN 0-306-44712-6

©1994 Plenum Press, New York
A Division of Plenum Publishing Corporation
233 Spring Street, New York, N.Y. 10013

Printed in the United States of America

PREFACE

The application of computational methods to solve scientific and pratical problems in genome research created a new interdisciplinary area that transcends boundaries traditionally separating genetics, biology, mathematics, physics, and computer science. Computers have been, of course, intensively used for many years in the field of life sciences, even before genome research started, to store and analyze DNA or proteins sequences, to explore and model the three-dimensional structure, the dynamics and the function of biopolymers, to compute genetic linkage or evolutionary processes etc. The rapid development of new molecular and genetic technologies, combined with ambitious goals to explore the structure and function of genomes of higher organisms, has generated, however, not only a huge and burgeoning body of data but also a new class of scientific questions. The nature and complexity of these questions will require, beyond establishing a new kind of alliance between experimental and theoretical disciplines, also the development of new generations both in computer software and hardware technologies, respectively.

New theoretical procedures, combined with powerful computational facilities, will substantially extend the horizon of problems that genome research can attack with success. Many of us still feel that computational models rationalizing experimental findings in genome research fulfil their promises more slowly than desired. There also is an uncertainity concerning the real position of a 'theoretical genome research' in the network of established disciplines integrating their efforts in this field. There seems to be an obvious parallel between the present situation of genome research and that of atomic physics at the end of the first quarter of our century. Advanced experimental techniques made it possible at that time to fill huge data catalogues with the frequencies of spectral lines; yet all attempts at an empirical systematization or classicaly founded explanation remained unsuccessful until a completely new conceptual framework, quantum mechanics, emerged that made sense of all the data.

The present situation of the life sciences is certainly more intricate due to the more amorphous nature of the field. One has to ask oneself whether it is a fair demand at all that genome research should reach the internal coherence of a unifying theoretical framework like theoretical physics or (to a smaller extent) chemistry. The fear seems to be, however, not unjustified that genomic data accumulation will get so far ahead of its assimilation into an appropriate theoretical framework that the data themselves might eventually prove an encumbrance in developing such new concepts. The aim of most of the computational methods presented in this volume is to improve upon this situation by trying to provide a bridge between experimental databases (information) on the one hand and theoretical concepts (biological and genetic knowledge) on the other one.

In the first two chapters of the book, Charles Cantor, Hans Lehrach, Richard Mott, and Günther Zehetner define the need for computer support in genome projects as seen from the point of view of experimentalists. Victor Markowitz, Richard Durbin, Jean-Thierry-Mieg, and Otto Ritter summarize in the next three chapters the current situation of genome databases and present some new concepts and prototypes for the next generation of integrated databases. Jaques Beckmann, Robert Robbins, and Tom Slezak et al. provide in the subsequent three chapters an overview of the techniques of genetic mapping and the representation of genomic maps in (integrated) databases. The next four chapters written by William Pearson, Eugene Myers, Jens Reich, and Gaston Gonnet will be devoted to the mathematical and statistical foundation of techniques to search nucleic acid and protein sequence databases, to recognize the significance of the occurring patterns, and to construct evolutionary trees with the help of such molecular sequences. William Taylor, Frank Hermann, Sándor Suhai, and Martin Reczko demonstrate the power of some new techniques (homology modelling, artificial neural networks, and genetic algorithms, respectively) in predicting the tertiary structure of gene products from their amino acid sequence and in identifying genomic signal sequence patterns. The final two chapters by Martin Bishop et al. and by Clara Frontali discuss statistical models of chromosome evolution and present some general ideas on deciphering the genetic message.

Most chapters of this volume have been presented as plenary lectures at the International Symposium on Computational Methods in Genome Research held on 1-4 July 1992 at Deutsches Krebsforschungszentrum (DKFZ) in Heidelberg. It is a great pleasure to thank here Professor Harald zur Hausen and the coworkers of DKFZ for their help and hospitality extended to the lecturers and participants during the meeting and to the Commission of the European Communities as represented by Bronwen Loder, Andreas Klepsch, and Karl Freese for the funding of the symposium. The organizers profited much of the help of the scientific committee of the symposium: Martin Bishop, Hans Lehrach, Bronwen Loder, Jens Reich, Otto Ritter, and Chris Sander. Furthermore, the editor is deeply indebted to Barbara Romanós, Monika Gai and Natascha Hess for their careful and hard work in preparing the camera ready form of the manuscripts.

Heidelberg, September 1993

Sándor Suhai

CONTENTS

CAN COMPUTATIONAL SCIENCE KEEP UP WITH EVOLVING TECHNOLOGY FOR GENOME MAPPING AND SEQUENCING?

Charles R. Cantor

Center for Advanced Biotechnology
6 Cummington Street
Boston, MA 02215
USA

I represent a consumer of genome informatics rather than a producer in many ways. I am not an expert in informatics, and yet I am going to make occasionally what I hope are some very provocative statements about informatics. So forgive me if I err. I started with a rather provocative title to this talk: "Can genome informatics keep up with the technology?" Let me add to that question as best I can right now from my contacts with computer industry: whatever wild projections we biologists may make about how much faster, how much better we will be 10 or 15 years from now in producing genetic data, maps, and sequence data - from the view point of computer hardware, there will be no difficulty on keeping up even with our most optimistic projections. From the view point of computer software, however, at least in the US, I think we mostly agree it has been a slow start, and it has not been very successful so far. There have not been many deliverables that the community can use. So my challenge to all of you is to make up your mind to catch up.

I do not think that there would be any fundamental problem in computer software keeping up with the projections I am going to give you for the genome project. What I do see as a significant problem in this area, and it may perhaps be even worse than in most areas, is that the human interface is the real limiting step. I mean by this both the interface between humans and computers, which is terribly frustrating at times and, probably even more so, the interface between humans and humans. There is much difficulty and miscommunication in our politics as well. I will try to describe for you in this talk, at least from my point of view, where the human genome project is going, and to point out where I can see some of those human challenges that I think need to be addressed. Besides introducing the human genome project, I will talk about its aims, show you some of the specific technological changes that I see coming over the next 10 or 15 years, and hopefully set the stage for Hans Lehrach's talk, which will follow, and then talk a bit about the specifics of the quantitative challenge that I see for the informatics.

Let me start with the genome project. I always like to think of it as characterized by three numbers shown in Table 1.

Of these three numbers, the estimate that there are 3 billion base pairs of DNA in the human genome, is a crude estimate. We do not really know that number very well; I fear it is low by maybe 10 or 20 %, but that remains to be seen. The only one of these numbers which is actually important is the number of genes, because the real purpose of the genome project is to find those genes and make them available so that biologists can study them. Over the last

Table 1. Three numbers characterizing the Human Genome Project.

The Human Genome

- 3 billion base pairs
- 24 chromosomes
- 100,000 genes

---

month or two, I have had groups of people frequently come to me and try to convince me that this number was too small, by a factor of 2 or more, or too large, by a factor of 2 or more. So this number is really quite uncertain, which is unfortunate.

We can see the whole genome at very low resolution by the technique of light microscopy. From a typical set of Giemsa-stained human chromosomes, somebody skilled in the art can tell if an individual is a male, because there will be a Y chromosome, and if an individual has a chromosome disorder such as Downs syndrome, because there would be 3 copies of human chromosome 21.

I estimate that there is something like about 600 bits of information in such a low resolution picture of the genome, and if the human genome project is successful, what we will be able to do some day is to view people at about 10^7 times higher resolution. A 10^7 increase in the amount of data tproduced by a test is quite significant, and it really poses quite a challenge both for the experimentalist to figure out how to span that range and for the information scientist, as well, who will have to analyze all that data. The challenge of this factor 10^7 is really less in the informatics field than it is for the experimentalist. The reason for this is that the current low resolution data is actually rather difficult to analyse, because it is an image rather than nucleotide sequence data which is, I think, much faster to analyze by standard computational methods. In taking this leap from looking at chromosomes to having the complete DNA sequence, we must proceed in several stages: no single method is available today that can cover the whole range of resolution. Currently, there are about 10 different types of maps of varying resolution that are available as one studies the genome.

The coarsest map currently, and I think destined to remain so, is the genetic map. One goal of the US human genome mapping effort to make a 2 million base pair resolution genetic map of the genome. There are various types of physical maps which directly describe DNA

Table 2. Different size scales of the Human Genome Project.

The Human Genome Project

The goals of the project:	
Complete a fine human genetic map	2 Mb resolution
Complete a physical map	0.1 Mb resolution
Acqire the genome as cloned DNA	5 kb resolution
Determine the complete sequence	1 bp resolution
Find all the genes	

Develop the tools to use the sequence for a variety of biological and medical applications

---

Table 3. The lengths of some characteristic DNA sequences.

GENOME SIZES

Largest continuous DNA sequence yeast chromosome 3	350 kb
E. coli genome (1 chromosome)	4.8 Mb
Largest yeast chromosome now mapped	5.8 Mb
Entire yeast genome	15 Mb
Largest continuous restriction map	20 Mb
Smallest human chromosome	50 Mb
Largest human chromosome	250 Mb
Entire human genome	3 Gb

structure and, of course, the ultimate physical map is the DNA sequence itself. Now, I need to point out to non-biologists that we must have this low resolution genetic map, though it is a very painful map to construct, as I will show you, because it is the only map that allows us to place a gene within the genome if all we know about that gene is a disease of a phenotype. It is the only map that lets us find disease genes and so we must have it. The other maps are much more straightforward exercises in molecular biology and physical chemistry.

What makes genome research challenging is the size scale over which these maps extend, from the single base pair resolution of the complete sequence to the many megabase resolution of current genetic maps (Table 2). The purpose of all this is, of course, to find the genes and, more and more, to develop the tools to use this information for biology and medicine. If there were a revolution in DNA technology, so that tomorrow we could determine the complete DNA sequence of a human in a single day, it would be very frustrating because our ability to use this information at the moment is rather slight. That ability is hindered for two reasons. At first, as I am sure you are aware and I will come back to this point later, our ability today to interpret raw DNA sequence in terms of a biological function is really fairly modest. We have to make major improvements in how we do that. But I think the most serious problem here is that the human is a lousy experimental organism. The classic genetic experiment that one wants to do, faced with one of these 100 000 genes of unknown function, is to knock that gene out, destroy it, and look at the consequences in an organism. That experiment is not only unethical in the human but it is impractical, because we are diploid and you have to knock out both copies of the gene and this is terribly hard. So we cannot work on the human and we cannot control breeding in the human, which is the way one has to do this experiment. It is, therefore, a necessary consequence not just a luxury that, as part of the human genome project, we study experimental organisms like the mouse which we hope are close enough to us humans, so that things make sense there, and we can do intelligent genetic manipulations.

We still have a long way to go to be able to sequence human DNA. Table 3 basically shows where we are coming from and where we have to go to. At the moment, the largest continuous DNA sequence just published as a result of a European collaboration is a length of 350,000 base pairs, one of the smallest of the known eukaryotic chromosomes, the yeast chromosomes. The largest maps actually completed are maps that cover simple genomes like *E. coli*, yeast and one arm of a human chromosome, containing about 40 megabases. This does not look bad at first glance, considering that the smallest human chromosome is 50 megabases and the largest is about 250 megabases. The problem is that, thus far, our experience has been that the difficulty in doing these projects does not scale linearly with the target size but more likely with the square of it. So it is a fairly steep learning curve that we have to accomplish, but I am pretty optimistic that at least at the present time the genome project is roughly on course. What I mean by that is shown on Table 4 which describes the original plan of the US genome project.

Table 4. The time chart of the Human Genome Project.

Years 1 to 5	1 % of the genome sequenced
	50 % of the genome mapped
	technology improved by 5 to 10 fold
Years 6 to 10	10 % of the genome sequenced
	100 % of the genome mapped
	technology improved by 5 to 10 fold
Years 11 to 15	100 % of the genome sequenced
	all of the genes found

Within the first 5 years, and our project is about 2 years old at the moment, there will be very little sequencing. Most efforts will have to be devoted to mapping and a major goal will be to improve our methodology of all types by something like an order of a magnitude. We need this improvement in order to allow us in the second 5 years to start intensive sequencing, presumably mostly genes, to finish the mapping and to try to gain another factor of ten in technology so that by the last 5 years of the project, all of the genome will be sequenced. As I think about it at the moment, and what I am going to try to convince you, is that this is far too pessimistic. I think we will do much better than this. Currently the emphasis of the genome project is almost all mapping. Let me start with genetic mapping and show you what this looks like in the human, where breeding is not possible. Figure 1 shows a hypothetical small family, with a mother and a father and 4 children. What human geneticists do is to look retrospectively at families like this and try to trace back what was inherited.

In the case in Figure 1, since each child inherits one copy of a genome from each parent, a child that gets the left chromosome from the father will have blue eyes (B) and Huntington's disease (HD). On the other hand, a child that gets the right chromosome from the father will have neither. Once in a while, about 1 % of the time, since these genes are a million base pairs away of each other, there may be a breaking and rejoining of the chromosomes before they will be passed to the offspring, and so a child may get blue eyes and not Huntington's disease or vice versa. By measuring the frequency of this rare event we are able to ascertain that these two genes are near each other in the genome and make some crude estimate of the distance between them. If you think about this for a moment, the problem is obvious: How many

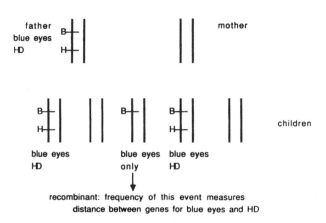

Figure 1. How genetic maps are made.

families do you know that have a hundred children? If you are looking for a 1 % effect you have to have 100 offspring. So, because our families are not that large, we have to pool data from different individuals, from different families in order to do human genetics. That makes the implicit assumption that the same trait or the same disease in different families is the same. Unfortunately, it is often not and that is why it is not uncommon to have major papers in human genetics being retracted. With existing technologies, especially in the human, it is extraordinarily difficult to find genes involved in polygenic disorders like the inherited susceptibility to heart disease, and other major targets of the genome project. For those people looking for good computer science projects to work on in context with genome projects: as far as I know, trying to deal with polygenic disorders is computationally very demanding by existing mapping methods. So it is a problem worth attacking.

The classical approach used in human genetics, to try to find genes one at a time, is exceedingly painful. Hence the genome project and the notion of making physical maps.

When physical mapping was first initiated, there were two basically rather different philosophies which were attempted by various groups. These are coming together now as I will show you later. In top down mapping we basically take a chromosome and divide it into non-overlapping segments by one method or another. The most frequent example is a restriction map, but there are other possibilities. We order these, and, if we want more information, we take them one at a time and again subdivide them into non-overlapping segments. This works very well; these maps can be completed, but they can provide only a low resolution view. The opposite approach, called bottom up, starts with the genome fragmented intentionally at random to guarantee that the pieces overlap. These fragments can be fingerprinted in one way or another and we look for clones that have an overlapping component. There are, however, very good statistical reasons that these bottom up maps cannot be completed.

Basically, by using one method or another we have now gained fairly extensive maps covering the major regions of a number of human chromosomes. Figure 2 shows one such map, just to make the point that once a map is complete, in this case 10 million base pairs starting from the telomere of the long arm of human chromosome 21, it is very easy to represent it; the informatics problems in dealing with it become essentially trivial. An interesting problem at the moment is, however, that we have different types of maps and in particular the genetic map and the physical map do not share a constant metric.

Figure 3 shows a larger section of human chromosome 21; on the left is the entire long arm as it looks like in the light microscope; in the centre is the genetic map; and on the right is the physical map. In order to get the genetic map on the same scale as the other two physical representations of the chromosome, one has to distort the genetic map considerably, because the genetic map is not a physical description of the structure. It is just measuring recombination frequencies and, for reasons that we do not understand, recombination is quite active near the very end of this particular chromosome and so the map gets rather distorted here. To summarize, we can make these maps, but it is painful; it really is boring and it stretches existing technology to the limit, so we really need more effective ways of doing this. What I want to show you is a very optimistic view of what is going on, to be able to increase the rate of mapping, and I think also the rate of DNA sequencing, tremendously over the next few years. There is, I suppose, a technological challenge for information science as well to try to keep up with these developments and to adapt to continuously changing methodology.

Let me start by showing in Figure 4 what I think is the modern view of mapping, quite different from the view of just a few years ago. Imagine a set of clones in many different libraries, or fragments, restriction fragments, radiation hybrids or whatever. Because the samples overlap, it is possible to fingerprint them by using probes. I will call them, here, sequences, but any probe, even just a synthetic piece of DNA can be used to look for overlapping clones in various libraries and fragments. The point is that we do this experimentally, and if we do not make too many mistakes, this will be a self-ordering process; these maps simply put themselves together. As long as we have a dense coverage of clones or

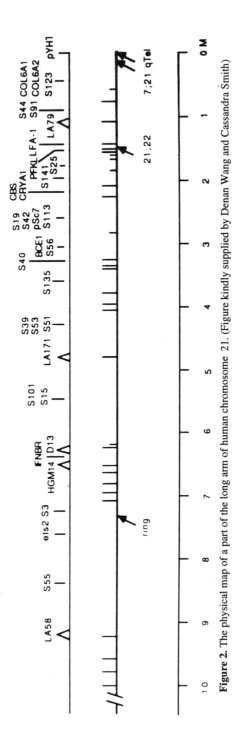

Figure 2. The physical map of a part of the long arm of human chromosome 21. (Figure kindly supplied by Denan Wang and Cassandra Smith)

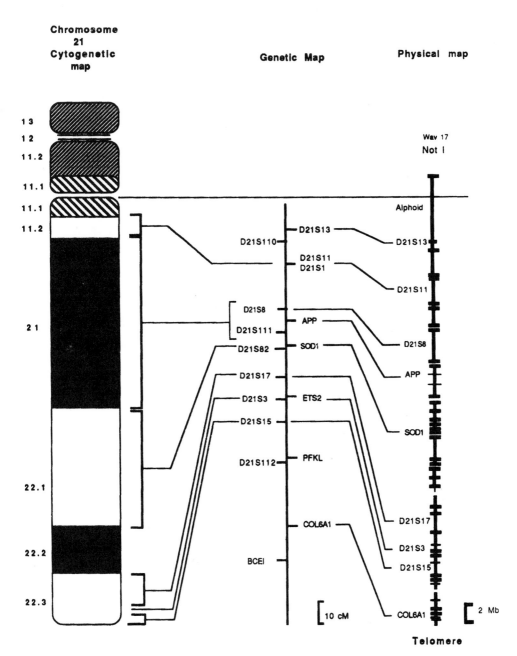

Figure 3. Genetic and physical maps of human chromosome 21. (Figure kindly supplied by Denan Wang and Cassandra Smith)

fragments, the maps assemble. It does not matter how we think about it, top down or bottom up, and it is very efficient. We can do it by hybridization, by fingerprinting, or by any other way. When we started the genome project, people were scared about so many different kinds of maps and asked themselves how will computer science be going to deal with all these different objects. The nice thing about the modern view from the viewpoint of informatics, that it does not make any difference; they are just part of the same picture.

7

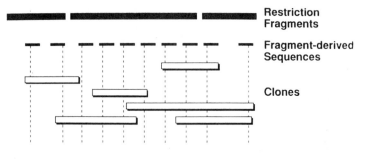

Matchup by hybridization or fingerprinting

Figure 4. Top down clone ordering.

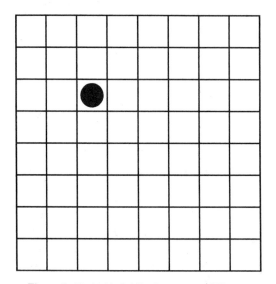

Figure 5. Typical hybridization to arrayed library.

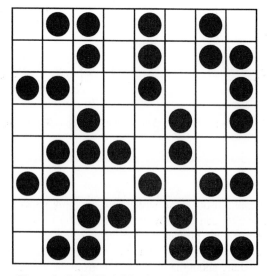

Figure 6. Optimal hybridization to an arrayed library.

The difficulty in doing clone and fragment ordering in the conventional way is that one handles these probes one at a time and looks at a large set of samples; there are typically 20,000 samples in an arrayed cosmid library. The sort of experimental result obtained is shown in Figure 5, which depicts, as an example, only a small number of samples out of the 20,000. Now, most molecular biologists doing this experiment will be looking for a gene in a library with a probe, and, getting this result they will congratulate themselves and say: "Is it not wonderful, I have now found a clone, a cosmid, a YAC, etc., that corresponds to my gene". From the view point of information science or from the viewpoint of large scale genome mapping, this is a very stupid experiment because it gains, from a tremendous potential amount of information, precisely one bit.

Instead, the experiment that we really want to do would light up, in an ideal case, 50 per cent of the clones, as shown in Figure 6. In fact, there is no other way to get information faster from this array in a single experiment. The question is how do we do this. The answer is: there are various ways. Instead of using one probe against an array of samples (probes and samples are equivalent here), we could have the individual samples be complex, they can be pools. Or we could have the probes be complex. Or we could be really ambitious and start with pools of probes and look for pools of samples. More details about this point will be given in the contribution of Hans Lehrach, but let me describe here just a very simple example for illustration. I will use our favorite organism, S. pombe, whose genome consists of only 15 million base pairs, and yet it has three relatively large chromosomes with an average size of 5 million base pairs. We can separate these chromosomes very easily, and we can separate pure fragments of the genome, so we can work with 10 % of the genome at once, or 5 % of the genome at once very easily.

What we do is we take this organism and separate either its chromosomes or pieces of those chromosomes by pulsed field gel electrophoresis and we use those chromosomes as complex probes, or we use pieces of those chromosomes as complex probes to fingerprint a cosmid library. In this case, the library has only a few thousand clones, because S.pombe is a small organism. The power of this approach is remarkable: in a single set of hybridizations we can assign several thousand clones to one of the three S. pombe chromosomes; so this kind of mapping can go very quickly. But the bad news is that hybridization is not black and white. If hybridization were perfect, we could accumulate data at an incredible rate. The problem that we have here is not a trivial problem for bioinformatics though it frequently occurs in molecular biology where people just use a threshold, throwing away most of the quantitative information. In reality, there is a tremendous amount of quantitative information in the intensive hybridization experiments if we had the proper algorithm to solve the problem. This imprecision in hybridization arises from cross hybridization. If we are using probes with 5 million base pairs complex, no wonder that there is some cross hybridization. If we use smaller probes, let us say only 500 000 base pairs, then the data is much closer to black and white; it is much easier to decide what will be positive and what will be negative.

I have described one way in which it is possible to speed up the process by pooling samples. Now, there is a second way. Until now, I have used the words probes and samples very carefully. For physical mapping, it makes no difference what kind of probe we are using as long as it gives us a nice hybridization pattern. For genetic mapping, what is required is to have to have a probe that is polymorphic in the population, i.e., we must have a DNA sample that is different in different people. Furthermore, we need to find such probes in very large quantities, ultimately tens or hundreds or thousands. In the past, this has been done largely by a random search that has been too inefficient. So we have been looking for ways to try to speed that up, and I want to show you, that we were capable of doing this by simple physical chemistry (Table 5).

It turns out that DNA sequences, which have all purines or all pyrimidines in a stretch, are extensively polymorphic in the genome, especially if they are regular repeats. We have developed an almost trivial trick to purify these sequences directly from a complex set of

Table 5. Sequence specific DNA triplexes.

Double strand
T C C T T T C T T T C
A G G A A A G A A A G

T C$^+$C$^+$ T T T C$^+$T T T C$^+$
Third strand

Stable at pH 5 to 6; unstable at pH 8 to 9

---

samples by forming triple strands with a complementary pyrimidine sequence. The key trick is that these triple strands form at slightly acidic pH, and they are unstable at alkaline pH. So just by changing the pH we can regulate whether they are strong or weak. Figure 7 shows how we implement this by using a magnetic bead that captures its target and allows to purify it away from the rest of the sample. Then we make the sample alkaline and throw the magnetic bead away. This whole process takes about an hour and the only thing that costs money is a tube and a magnet.

Table 6 demonstrates this procedure by a model experiment. We took a large number of DNA molecules, 99.5 % of which do not contain a target for triple strand formation and mixed them with 0.5% (50,000 clones altogether) of a sample which did contain a target for triple strand formation, $(TC)_{45}$. We subjected this mixture to one round of magnetic enrichment and

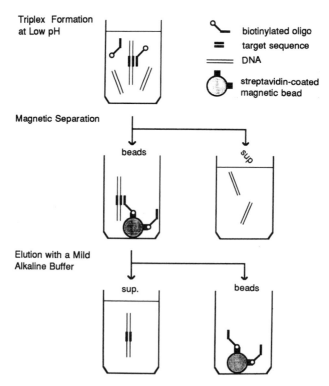

Figure 7. DNA triplex formation and magnetic separation (Figure kindly provided by Takashi Ito).

Table 6. Triplex-mediated enrichment of target plasmids from a reconstructed library (Adapted from Ito et al. PNAS, 89, 495-498, 1992).

	colonies	
	white (pTC_{45})	blue (pUC19)
before enrichment		
number	5.0×10^4	1.1×10^7
(%)	(0.5)	(99.5)
after enrichment		
number	4.0×10^4	0.5×10^2
(%)	(99.9)	(0.1)

then recloned and found that 99.9 % of the new library was the desired material with only 0.1 % contamination at that point. Our yield was 80%, with a purification factor of 140,000 in one step. If this is not good enough, we can do it again. The point is that we can obtain useful probes faster than anybody could possibly study them, so available information gathering is going to be quicker.

What I would really like to do is to find a way of speeding up all of these mapping processes by an additional factor of 50. We are experimenting with a way that might work, though it does not completely work yet. But I want at least, to show you the thinking. The rationale for how to gain a factor of 50 in speed is to use multiple colors. The experiments I have shown up till now, and in fact all mapping, were just black and white. Instead of using just one label, if we had 50 different labels, then we could study 50 different samples mixed together at once, and everything would go 50 times faster. If we want to do this, we have to have a way of attaching many different labels to biological molecules, and Table 7 shows how we are trying to do that.

We use the fact that one can put biotin on nucleic acids anywhere, and we can capture that biotin with streptavidin, a protein that is incredibly stable and binds biotin with an affinity of 10_{15}. We recently produced in our laboratory a chimeric protein fusion between streptavidin and the protein metallothionein. Streptavidin is a tetramer, metallothionein is a

Table 7. Streptavidin-metallothionein chimera (Table kindly provided by Takeshi Sano).

Expression vector:	pTSAMT-2
Amino acid residues:	190 per subunit
Cysteine residues:	20 per subunit
Subunits:	4 (subunit tetramer)
Molecular mass:	19.5 kDa per subunit
	78 kDa per molecule
Biotin binding:	4 per molecule
	1 per subunit
Metal binding:	$6.7 + 1.0$ Cd^{2+} per subunit
	(X-ray fluorescence)

Table 8. Metal binding properties of metallothionein (Table prepared by Takeshi Sano).

Zn, Cd, Hg, Pb, Bi, Sn, Co, Ni, Bi, Fe, (TcO) Seven ions per molecule	Cu, Ag, Au Twelve ions per molecule

Stability: Zn (II) < Pb (II) < Cd (II) < Cu (I), Ag (I), Hg (II), Bi (III)

$$K'_{Cu} = 10^{-17} - 10^{-19}\ M$$
$$K'_{Cd} = 10^{-15} - 10^{-17}\ M$$
$$K'_{Zn} = 10^{-11} - 10^{-14}\ M$$

Table 9. Stable metal isotopes that can be bound by metallothionein (Table prepared by Takeshi Sano).

Stable Metal Isotopes That Can Be Bound by Metallothionein

$_{26}$Fe	^{54}Fe ^{56}Fe ^{57}Fe ^{58}Fe	$_{50}$Sn	^{112}Sn ^{114}Sn ^{115}Sn ^{116}Sn ^{117}Sn ^{118}Sn ^{119}Sn ^{120}Sn ^{122}Sn ^{124}Sn
$_{27}$Co	^{59}Co		
$_{28}$Ni	^{58}Ni ^{60}Ni ^{61}Ni ^{62}Ni ^{64}Ni	$_{79}$Au	^{197}Au
$_{29}$Cu	^{63}Cu ^{65}Cu	$_{80}$Hg	^{196}Hg ^{198}Hg ^{199}Hg ^{200}Hg ^{201}Hg ^{202}Hg ^{204}Hg
$_{30}$Zn	^{64}Zn ^{66}Zn ^{67}Zn ^{68}Zn ^{70}Zn		
$_{47}$Ag	^{107}Ag ^{109}Ag	$_{82}$Pb	^{204}Pb ^{206}Pb ^{207}Pb ^{208}Pb
$_{48}$Cd	^{106}Cd ^{108}Cd ^{110}Cd ^{111}Cd ^{112}Cd ^{113}Cd ^{114}Cd ^{116}Cd	$_{83}$Bi	^{209}Bi

Total 50 species

monomer, but in the structure the streptavidin is dominant, so that the overall structure is a tetramer. It binds four biotins and each metallothionein subunit binds 7 metals, so the tetramer binds 28 divalent metals (Table 8). So we can put 28 metals anywhere there is a biotin, and we can put a biotin anywhere there is a nucleotide base. So we can basically turn nucleic acids almost into metals, if we want to. Now, what is attractive about this scheme is that metallothionein binds lots of different metals. Furthermore, the binding constants for the most part are extremely tight. Once these metals are bound, they are not going away again, and there are many different ways to detect them, the most attractive one at the moment being mass spectroscopy.

The advantage offered by the use of metallothionein is that there exist some 50 different stable metal isotopes (Table 9) which can now be used to tag nucleic acids, and these can all be detected relatively easily by mass spectrometry. The resolution required is only basically one mass unit, which is trivial. The main difficulty at the moment is that the equipment is very expensive. I hope this will change. So, I have shown the number of different ways in which, I think, things may get substantially better. Now, let me turn from mapping to DNA sequencing, because most of the effort in all of these projects is going to be DNA sequencing, and, I think, most of the challenge really is to deal with the DNA sequence data.Table 10 shows the general scheme by which, at least today, all DNA sequence is obtained.

Table 10. General scheme for DNA sequencing.

---

end label (*)
*ATCCGAGGACACTATTACT

partial cleavage at one base (or the equivalent) : A	length
*A	1
*ATCCGA	6
*ATCCGACCA	9
*ATCCGAGGACA	11
*ATCCGAGGACACTA	14
*ATCCGAGGACACTATTA	17

separate by length with resolution finer than one base
lengths observed via end label = cleavage positions
repeat for three other bases: T,C,G

Table 11. New DNA sequencing modalities.

Project Areas:

Ultrasensitive DNA detection in gels
Improved DNA sequencing chemistry/enzymology
Multiplexing strategies: many colors or many serial probings
Capillary electrophoresis
Sequencing by hybridization
Scanning tip microscopies (AFM/STM/etc.)
Flow cytometric sequencing

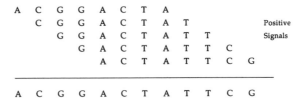

```
A   C   G   G   A   C   T   A
    C   G   G   A   C   T   A   T                       Positive
        G   G   A   C   T   A   T   T                   Signals
            G   A   C   T   A   T   T   C
                A   C   T   A   T   T   C   G
─────────────────────────────────────────────

A   C   G   G   A   C   T   A   T   T   C   G
```

Figure 8: The basic idea of DNA sequencing by overlapping oligomers.

Current sequencing is really based on two principles. First, we need a way of labelling one end of the molecule. That could be done with a radioisotope, a fluorescent label, a mass label as we have seen above, or any kind of label we can think of. We can selectively fragment the DNA molecule or synthesize it in a base-specific way to generate fragments whose lengths tell us about the location of the bases. That is all, and it is the way all sequencing is currently done. The difficulty with this approach is that it is intrinsically slow, because we have to do a physical separation to detect the fragment lengths, and physical separations are usually slow. In addition, also it is a serial analysis and, as such, it is very inefficient. There are many schemes to try to improve this: Table 11 shows a list of some of them.

Some of these methods may be problematic to work with at the moment. I think the one that has the best chance to impact almost immediately on the human genome project is sequencing by hybridization where we basically determine all n-tuple words in the sequence and, because there is an overlap, we can assemble them into the complete sequence. The basic idea of this procedure is shown in Figure 8.

The problem of assembling words into sequences is computationally not a trivial one. There are some very interesting questions when we have repeating sequences where we get branch points and we obtain a non unique assembly. But the beautiful thing about this approach is that by simply taking our unknown sample and determining its content by hybridization of an oligonucleotide to the sample, it allows us to process enormous numbers of samples in parallel. We can take an array with 20,000 samples, probe it with one of the oligonucleotides and we will be determining the sequence content of a particular word in all the 20,000 samples simultaneously. And, of course, we can pool some of these in an intelligent way and gain information even faster. Some people talk about making a chip in which we could lay down all 64,000 possible octanucleotides and interrogate the chip with the sample to determine its complete octanucleotide content in a single experiment. It is just the costs of making that chip which inhibits this experiment at the present time. I am absolutely convinced, because I have known a number of people who have been attracted to work in this area, that in less than five years somebody will be building these chips and we will be doing sequencing this way for sure. How accurate it is going to be remains to be seen and the exact formulation of what will be developed remains to be seen as well.

When I mention the word chip I always think, of course, of an electronic chip, and, in the most extreme version of how this might be done, hybridization would actually change the property of some electrical device and give a direct signal. Now, chips like this have already been built. They are not very good yet, but I think we will make progress. If we will be able to do a massively parallel analysis with hybridization sequencing and in mapping by pools of probes and pools of samples, the speed by which we will be able to obtain genome information is going to increase by a factor of somewhere between 100 and 10,000.

Let me give here some examples that are all hypothetical and, perhaps, widely optimistic. If hybridization were perfect and we tried to make a genetic map by having the entire pedigree on a blot and by having pools of 250 probes at the time, we will need 9 colors if we had 9 pools to look at them all simultaneously (Table 12). If we go through the calculation, in a single hybridization (less than one day's work), we will be able to find the location of a gene

Table 12. One hybridization tests linkage at 1 cM resolution.

If HYBRIDIZATION WERE PERFECT

6 cM map is 500 probes

Blot:	entire pedigree
Pools:	9 pools of 250 each (binary sieve)
Colors:	9

1 hybridization tests linkage at cM resolution.

Table 13. Ordering of the test clone library by hybridization.

If HYBRIDIZATION WERE PERFECT

Filter:	2×10^4 samples = single chromosome cosmids (7x coverage)
Probes:	$10^4 = 3$ hits per tiling path cosmid
Pools:	10 sets of 10 pools of 500 each (binary sieve)
Colors:	20

5 hybridizations would order the library.

Table 14. Speed estimate of DNA sequencing by hybridization.

If HYBRIDIZATION WERE PERFECT

Filter:	2×10^4 samples; 200 bp each
Probes:	6.4×10^4 octanucleotides
Pools:	20 - 24 pools, 3.2×10^4 to 8×10^3 probes each
	(great redundancy over binary sieve)
Colors:	26 including controls

1 hybridization determines sequence except for branch points;
4×10^6 bp per hybridization; up to 10^8 bp per day.

Table 15. Estimated DNA sequencing speeds from several groups.

	RAW KB PER PERSON PER DAY	
	1993	1996
Church	80 - 240	4800
Gesteland	17	---
Smith	54	200 - 300
Mathies	140	500
Jacobson	---	500
Drmanac/Crkvenjakov	300	1000 - 2000
"average"	118	833

---

anywhere in the genome, in contrast to about 2 years of work at the present time with existing methods. We might even do much better; I think we may have as many as 50 colors eventually, so talking about 9 colors is conservative. Physical mapping is much more difficult. If we imagine having a sample of 20,000 clones, which is seven times the coverage for a chromosome divided into cosmids with each 50,000 base pairs in size, and having 10,000 probes (which is reasonably dense but perhaps not quite dense enough) and if we construct the appropriate set of pools, in this case 10 sets of 10 pools of 500 samples each, 20 colors, 5 hybridizations would order the library (Table 13).

To do this kind of ordering at the moment by standard methods takes a group of 4 people about 3 or 4 years. So potentially we can gain a lot here.

What is the prospect in DNA sequencing? Table 14 shows that this is a much more complex problem. Taking again as an example 20,000 samples and trying to sequence 200 base pairs on each sample with octanucleotide probes, making pools (in this case with a great redundancy over a binary sieve) and using 26 colors, a single hybridization would determine about 4 million base pairs of DNA sequence. Since we might be able to do many of these a day, we could approach sequencing rates of about 10^8 base pairs a day.

In Table 14, I present some of my own estimates of how fast people will be able to sequence in the future. In Table 15, other people's estimates are shown. I took a survey about a year ago of some of the people trying to develop high speed sequencing methods, and I asked them how fast they will probably be able to sequence, in this case in raw kilo bases per person per day, in 1993 and 1996. The average of the different methods was 100,000 base pairs a day in 1993. Actually, Wilhelm Ansorge at EMBL has essentially reached that rate already, so in a way this is too conservative, and something like a million base pairs a day in 1996 does seem to be within our reach. I hope those numbers are beginning to either attract you or to frighten you, because I think they are realistic.

Current methods deliver 20,000 bases of raw DNA sequence a day. We are pretty sure that they can be optimized by a factor of 10. The mass spectroscopists are pretty optimistic that they can go another factor of 100 over that, and sequencing by hybridization is approaching to 10^6 or 10^8 base pairs per day. Those numbers are not ridiculous because let me remind you that when *E. coli* replicates, it sequences its own DNA at the rate of 2×10^8 base pairs per day. Let us suppose that some laboratory will be sequencing at the rate of 10^8 base pairs a day five years from now. There are two issues to be discussed: First, do we really need those rates? Yes, we do, I will try to convince you in a moment. But then the most serious issue I think is: If we had those rates what will we do with the data? A speed of 10^8 base pairs a day will produce actually several telephone books worth of information that we will get every morning on our desk. How shall we able to analyze it, how shall we even think about it? This is not only a question of having the software to reassemble or to check it against the databases. We would need to produce every morning a sort of DNA sequence newspaper with headlines and columns and other things like this in it, so that, someone who is only interested in the cricket scores could go to that section of the newspaper and ignore all the rest. I do not know how we are going to do this, but I think we need to do it, and the need to be able to do it quickly is really quite pressing, because DNA sequencing rates are going to grow tremendously.

Let us return to the present. Today, none of these fancy techniques are available. At the moment we have problems with all our methods: they are imperfect, they are slow, they have noise. Still, in spite of the fact that today's methods are rather primitive, the situation right now is that, with the existing methods, almost any laboratory in mapping or sequencing is producing the data faster than it can analyze it. And, as we have seen before, we are going to increase data production by a factor of 10^4, so there is really quite a demand on information science to help us keep up with it. Now, how much information do we have to deal with? We might think of the human genome of a single person (Table 16) as filling up 200 telephone books each with 1,000 pages books. Or by another calculation, if we would read that sequence at a standard reading speed, day and night, 7 days a week, it would take 26 years. For most

GENOME/PHONE BOOK ANALOGY

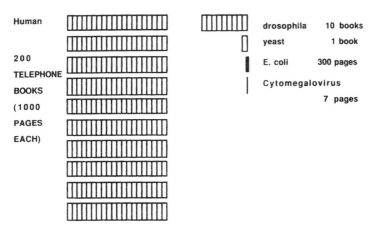

Figure 9. Typical genome sizes.

other data sets, we never do read the whole data; we do not perform a global search. But in genome research, we do so frequently; we have to analyze the entire data set, so this is a very challenging problem.

The real situation is, however, much more interesting. It would be quite a modest task just looking at one human. But in genome research, we are comparing people. That basically is the genetic experiment we have to do on the computer. I would like to imagine that our technology will become good enough to enable us to sequence everybody. When you are born you are sequenced, and if we are clever and compress the data down, so that we have a database just of differences between people and throw away all the stuff that is the same, and take the extremely conservative estimate that we differ from each other at the level of 0.1 %, that means 3 million differences per any individual. If we assume an (underestimated) earth population of 4 billion individuals, the resulting database will be required to hold (just handling the differences between humans) information about 10^{16} base pairs (Table 16).

Now, by existing computer technology, a single human genome is small stuff, but 10^{16} is a very big number. It means about a million CD ROMs, and I think there is a reasonable likelihood that we will, 15 years from now, be thinking about having to deal with a database of that magnitude and trying to get it. The reason is that this information is going to be tremendously useful in clinical diagnostics. Somebody might walk into a clinic somewhere in the world with a peculiar set of symptoms and a corresponding DNA sequence and, by linking those up in a common database, a doctor will be able to give a more accurate diagnosis, more accurate therapy than we could imagine otherwise. Let me illustrate this point by the example of Duchenne Muscular Dystrophy (DMD).

Table 16. Human genetic diversity: the ultimate human genetic database.

- Any two individuals: differ in about 6×10^6 bases (0.1 %)
- On the earth at present there are about 5×10^9 individuals
- A catalogue of all differences would require 3×10^{16} entries
- This catalog may be needed to find the rarest, or most complex, disease genes

17

Currently, a world-wide pilot study is in effect to test a DNA kit for the analysis of DMD. This is not yet in total clinical use, this is a set of trials. This group is gathering DNA data on a very serious disease and trying to integrate it. If we first look at this we might say: "My goodness, how impressive!" But I think we can easily imagine that, as it becomes a very large scale operation world-wide and clinical data is being gathered all over and integrated, the political and organizational problems of managing this kind of network are going to be absolutely horrendous. First, the mountain of data is actually very large, but in addition one may actually want to be able to look, not just at the DNA sequence of two patients who seem to be similar, but also compare their x-ray data, and this is not so simple. Furthermore, because the constraints of medical systems and information handling in different countries are so different, in some cases the political problems to try to integrate this into a network are very serious.

In summary, I have tried to show that there are tremendous opportunities here but there are also tremendous challenges, and at the heart of it, I think, lies the difficulty in communication between biologists and computer scientists: a problem which is just getting exaggerated as they try to live closer together. I hope that problem will solve itself just by more familiarity with each other's terms and limitations, but I think we are going to be faced with lots of peculiar misunderstandings as a war or clashes of cultures as the project proceeds. I want to close this by telling one example, which is only two days old and therefore very fresh in my mind and very frustrating, of how such misunderstandings can lead to great difficulties. The example comes from Tom Jukes, a molecular evolutionist at Berkeley, who recently submitted a grant to do what he has been doing for twenty years: comparing sequences. He is interested in forms of molecular evolution, codon bias and codon drift, and he goes to the databases in a global fashion, like many of us do, looking for patterns across organisms. He submitted this grant to the usual grant channels and received a query back: "Are you going to work with any human sequences?" He said: "Of course I do, that is part of the database!" Then he was told that he had to get human subjects' approval in order to interrogate the database because the database has human sequences in it. He protested loudly but ultimately he capitulated and this grant was processed for human subjects' approval. That was only a local University of California committee, not a national issue. But if this kind of nonsense becomes a national or international policy, it will become a nightmare to think about DNA.

INFORMATICS AND EXPERIMENTS FOR THE HUMAN GENOME PROJECT

Hans Lehrach, Richard Mott and Günther Zehetner

Imperial Cancer Research Fund
Lincoln's Inn Fields,
London, WC2A 3PX
UK

SUMMARY

The understanding of the information stored in the mammalian genome is the major challenge to biological research for the next decade, requiring close co-ordination of work in many laboratories and major progress in the capability of generating, analysing and storing very large amounts of data. In many respects analogous to developments in high energy physics, the genome project will require a high degree of innovation in data acquisition, data storage, and data analysis.

In our work on the analysis of the information content of genomes of different organisms, we have developed a number of experimental and informatics tools and concepts. These are based on considering experimental materials (clones, cDNA populations) as objects. Experiments either transform objects into objects of a different type, or establish experimental relations between objects. This allows a close integration between experimental work and the relational database and logical programming concepts used to describe the experiments and their results.

INTRODUCTION

Genomes of mammals contain a large amount of information (approximately 3 billion bases in the case of the human genome) which the organism uses during its development and continuing existence. The information stored in this base sequence encodes not only the machinery that each cell requires to carry out tasks like generating and using energy, but also the information for constructing this machinery, the codes used in the construction, as well as the information controlling the selective information readout. The task of deciphering this information therefore goes far beyond carrying out the experimental work necessary to determine the DNA sequence of the genome, but will also have to include the acquisition and analysis of a large amount of additional information necessary to understand the code used in its readout. In a sense, this process can be considered to be analogous to the reconstruction of the design of a workstation, its machine language, its operating system, its programming languages and its programs from the bit pattern found on its disk.

This can realistically only be attempted if not only the genomic sequencing information is generated, but if also large amounts of additional information are available on how these sequences are read out into mRNAs and proteins in different tissues and stages of development.

DATA VOLUMES GENERATED UNDER THE HUMAN GENOME PROJECT

Since the information content of the genomic sequence already approaches one gigabyte, we can expect, that even the genomic sequencing component of the genome project will have to analyse tens to hundreds of gigabytes of raw sequence data (in turn extracted from hundreds of terabytes of scanning data). Similarly, analysis of the expression patterns of cDNAs or proteins in different materials will be likely to require acquisition of data of the same order of magnitude. It is therefore probable that the genome project will require the efficient generation, analysis and co-ordination of terabytes of data, generated by many different approaches in an international collaboration.

These values are considerably larger if the volume of image files generated in acquisition of the primary data is considered. Even larger amounts of data will have to be generated if genomes of many different species or individuals are to be analysed (hundreds to thousands of terabytes). Although such amounts of data are well within the range of those generated and analysed by large physics or space research projects, they exceed by far the amounts encountered currently in biological research, necessitating appropriate change in the concepts and approaches used in data acquisition, storage and analysis.

THE OBJECT-RELATION MODEL OF EXPERIMENTAL ANALYSIS

The approach to allow such large scale data generation and analysis which we have chosen is based on treating experimental work as establishing relations between experimental materials (experimental objects). These experimental objects are often clones from specific genomic (YAC, cosmid, P1, exon or cDNA) libraries, but can also include other materials (organisms, tissues, nucleic acid or protein preparations from tissues or entire organisms, protein spots in e.g. 2 D gel electrophoresis), as well as information objects (images, sequences). Objects are usually generated from other objects by well-defined experimental procedures, which can range from a mating experiments, which generates new objects (mice) from previous objects (mice), to library construction in which e.g. a collection of YAC clone objects (a YAC library) as generated from a nucleic acid preparation, which in turn is ultimately derived from an organism. Similarly, individual objects (e.g. clones) can be transformed into other objects for example in the derivation of a DNA preparation 'object' from a clone 'object'. This derivation establishes a relation by descent and transfers information (e.g. the position on the genome) to the newly generated object. If therefore either the original clone or any material derived from it are mapped, the position of the unmapped object is determined.

Similarly, a subclone will share features like its position on the physical or genetic map with the original clone. If either subclone or original clone are mapped, this will therefore determine the localisation of both objects. Analogously the derivation of cDNA clones from mRNA transfer the information that the cloned sequences are likely to represent transcripts in the tissue used as source for the RNA to the newly created clones.

This type of relation by descent is complemented by relations established by experimental results. If, for example, a cDNA clone is hybridised to a filter grid containing YAC clones, a positive hybridisation result will establish a relation between that cDNA clone and the genomic clone. Within the uncertainty given by possible cross hybridisation, this assigns the cDNA clone and all hybridising genomic clones to the same position in the genome, identifies overlaps between the genomic clones, and establishes the presence of a transcript in the region of the genome.

TRANSIENT AND PERMANENT OBJECTS

Objects can be either permanent (e.g. the clones from a library picked into microtitre plates) or can be virtual, to be regenerated again, when needed. Regeneration of objects is, however, only possible for objects which can be regenerated deterministically to give reproducibly the same material (e.g. a probe made from a specific DNA). Any derivation like random subcloning will not regenerate exactly the same objects, requiring either storage of the original object, or the partial or complete characterisation of the new objects (e.g. random subclones) to eliminate the uncertainty introduced due to the random nature of the derivation step.

To convert random library clones into permanent and therefore also shareable objects, clones have to be picked into microtitre plates. Libraries of such clones can be replicated, distributed and used in many different types of experiments. Since the number of objects (clones) needed to represent the entire genome and most of the transcripts occurring in different tissues and different stages of development is likely to be of the order of millions, the number of potential pairwise relations of the form "clone A hybridises to clone B" is quite staggering. It seems therefore essential to use experimental strategies which are able to probe for a large number of possible relations of this type in parallel. We have therefore concentrated on the development of techniques able to analyse large numbers of clones in parallel, made possible by the use of automated equipment to transfer clones or DNA from the microtitre plate pattern used in storage or higher density equivalents to the very high densities (tens of thousands of colonies or DNA spots per 22x22 cm filter) needed for efficient hybridisation experiments.

THE USE OF SHARED EXPERIMENTAL OBJECTS: THE REFERENCE LIBRARY SYSTEM

The concept of shared experimental objects, for example in the form of clone libraries stored in microtitre plates made available to other laboratories, automatically allows sharing of experimental tasks, a concept underlying the ICRF Reference Library System. If the same clones are available for experimental work in other laboratories, and a common nomenclature is used, experimental results determined by many laboratories can be combined, allowing a drastic reduction in the overall effort, and access to a much wider range of techniques than available in any single laboratory.

THE DESCRIPTION OF RELATIONS BETWEEN OBJECTS CAN FORM THE BASIS OF A RIGOROUS DESCRIPTION OF EXPERIMENTAL OBJECTS AND RELATIONS

To allow a rigorous description of experimental results based on the model of establishing relations between objects, both objects and experiments will have to be rigorously defined. This is most easily assured by the distribution and use of common objects which is obviously quite easy in the case of clones. This can also be achieved by the use of sequenced tagged sites, [1] based on the provision of primer sequences allowing (within limits) the deterministic regeneration of specific short DNA sequences from sequence information on primers for PCR reactions. Provided the objects and experiments are rigorously defined and the objects are distributed together with any associated data obtained from previous experiments, then further manual or automated experiments/analysis can be carried out irrespective of the time and place of generating the original data. In addition, new data can be interpreted together with all pre-existing information, simplifying the resolution of the large number of conflicting results likely to be generated over time.

This concept should therefore allow a complete description of the experimental work, and should offer the maximal possibility of combining different data sets without loss of information. Generation of such a complete data set in work distributed over many laboratories can be further improved if experiments can be entered on-line in a distributed notebook system, allowing the direct entry of information on objects and experiments into a shared, online database, while the experiments are being carried out. The development of such a notebook database, most useful in e.g. large scale mapping or sequencing projects requiring the analysis of many different objects by well characterised techniques, is in progress.

MATCH BETWEEN EXPERIMENT AND LOGICAL PROGRAMMING CONCEPTS

The concept of establishing relations between objects has an obvious match to the type of information stored in relational databases. An even more flexible system of storing and manipulating such relations is, however, provided by logic programming techniques. This match between experimental strategies and logic programming is most obvious in the definition of programs in Prolog as "declaring some facts about objects, and their relationships; defining rules about objects and their relationships; and asking questions about objects and their relationships" ("Programming in Prolog", Clocksin and Mellish, page 2). This relation together with some rules about the interpretation of those relations can then be used to establish the wide range of different information about the object. For example the detection of hybridisation between a probe and a clone will correspond to setting the unknown genomic positions of clone and probe to the same initially unknown value. As soon as either clone or probe are localised, both positions can be considered to have become instantiated to the same value.

AUTOMATION AND EXPERIMENTAL STRATEGIES

Spotting robot

To achieve a high throughput of data at relatively low cost, we have therefore developed robotic systems which allow the construction of membrane filters carrying much higher spot density than can be achieved manually, or by the simple mechanical devices developed for this purpose. Such filters carrying up to 36000 clones or DNA spots per 22x22 cm filter form the basis of much of our work, as well as the distribution of cosmid and YAC filter grids through the ICRF Reference Library System.

Starting from a first purpose built robot approximately five years ago, we have now developed and started to use a third generation system, based on linear- drive technology, which should provide sufficient capacity for the medium term future.

Though the production of such high-density membranes has been the most critical step in the automation of the entire process of data generation, other steps in the procedure of library analysis have also become sufficiently demanding to warrant the development of appropriate technology, since each improvement in the capacity at a critical step tends to shift the bottleneck to a different step in the process, which in turn have to be eliminated.

Development of special, high-density microtitre plate analogues

An essential step in this work has been the development of a new, higher density microtitre plate format, which essentially contains the equivalent of four standard microtitre plates in one plate. This has critically increased the number of clones, which can be stored in each -70°C freezer, and allowed the handling of clones in multiples of 384 which has led to a four fold improvement in the speed of library replication and library spotting.

Clone picking robot

Similarly, clone picking has become more and more a limiting step. A robotic device capable of automated picking of tens of thousands of clones per day has been constructed (Ahmadi and Curtis, unpublished).

PCR robot

As an alternative to the spotting of organisms followed by growth of colonies on the plate and a filter processing step, we have recently also made essential progress in developing of spotting DNAs, based on the large scale amplification of DNA samples in PCR. For this purpose a machine able to amplify up to 48000 samples in parallel has been built, and is being tested at the moment. Such a device will also play an essential role in the development of sequencing by hybridisation [2, 3] based on hybridisation of short oligonucleotides to large numbers of clones in parallel.

LIBRARY DEVELOPMENT

Genomic libraries

This system has been applied to a number of different types of libraries. Libraries constructed include cosmid libraries from chromosomes X, 1, 6, 7, 8, 11, 13, 17, 18, 21 and 22 [4], (Nizetic et al., unpublished). All libraries are available as high-density filter grids through the ICRF Reference Library System. Total genomic libraries have been constructed in the cloning vector P1 with an average insert size of 90 kb, corresponding to approximately 1.5 fold genome coverage (Francis, unpublished), as well as YAC libraries of the human (currently 8 fold coverage, 630 kb average insert size) and mouse (4 fold coverage, 700 kb average insert size) genomes [5], (Schalkwyk et al., unpublished; Maier-Ewert et al., unpublished). Large scale distribution of YAC filter grids has started recently, while P1 filters have been distributed to a few outside laboratories.

Gene libraries

Distribution of filters of cDNA libraries from different tissues will also allow the application of hybridisation technique to establish relations between clones. Currently available are a human brain cDNA library of approximately 40,000 clones, a mouse brain cDNA library of 20,000 clones and a pancreas tumour library of 20,000 clones. Additional libraries are under construction. Based on these resources, we are therefore able to establish large numbers of relations between clones based on the observation of positive hybridisation of specific probes to the high-density filter grids of the corresponding libraries.

APPLICATION TO PHYSICAL MAPPING

To test some of these concepts and to provide the groundwork for analogous work on the analysis of the human and mouse genomes, we have established a complete physical map of the genome of the fission yeast *S. Pombe* in YAC, P1 and cosmid clones [6, 7, 8]. Work on the human chromosomes, based on similar techniques, has been started.

INTEGRATION OF PHYSICAL MAPPING, TRANSCRIPT AND SEQUENCE INFORMATION

This physical mapping information can be integrated with genetic or transcript information, if clones are used to hybridise to YAC, P1 or cosmid clones. This information can then be combined with partial sequence information on cDNA clones and exon trap clones, established by hybridisation of high-density filter grids carrying DNA from cDNA clones to short oligonucleotides [3]. The information can then be used to establish a complete catalogue of transcribed sequences, of the clones in these libraries.

SUMMARY

The very large amount of data which have to be generated during the analysis of the human genome, as well as any other large genome, are in the process of forcing a major change in experimental strategies and informatic tools. The object-relation model of experimental work developed on the basis of the ICRF Reference Library System seems well suited to aid in the data acquisition, storage and analysis required for this project.

ACKNOWLEDGEMENTS

I would like to thank all members of my group for discussions and experimental results. In addition, I would like to especially thank Otto Ritter, Sándor Suhai, Shaun Cutts, Steve Bryant, Martin Bishop and Francis Rysavy, for many discussions under the Integrated Genome DataBase (IGD) project, which have played a major role in formalising the concepts described here.

REFERENCES

1. Olson M., Hood L., Cantor C. and Botstein D. (1989). A common language for physical mapping of the human genome. *Science* 245: 1434-1435
2. Drmanac R., Labat I., Brukner I. and Crkvenjakov R. (1986). Sequencing of megabase plus DNA by hybridisation: Theory of the method. *Genomics* 4: 114-128
3. Lehrach H., Drmanac R., Hoheisel J., Larin Z, Lennon G., Nizetic D., Monaco A.P., Zehetner G. and Poustka A. (1990). Hybridisation fingerprinting in genome mapping and sequencing.*Genome Analysis* 1, Cold Spring Harbor Laboratory Press
4. Nizetic D., Zehetner G., Monaco A.P., Gellen L., Young B.D. and Lehrach H. (1991). Construction, arraying, and high-density screening of large insert libraries of human chromosomes X and 21: their potential use as reference libraries. *Proc. Natl. Acad. Sci. USA* 88: 3233-3237
5. Larin Z., Monaco A.P. and Lehrach H. (1991) Yeast artificial chromosome libraries containing large inserts from mouse and human DNA. *Proc. Natl. Acad. Sci. USA* 88: 4123-4127
6. Maier E., Hoheisel J.D., McCarthy L., Mott R., Grigoriev A.V., Monaco A.P., Larin Z.and Lehrach H. (1992). Yeast artificial chromosomes clones completely spanning the genome of *Schizosaccharomyces pombe. Nature Genetics* 1: 273-277
7. Hoheisel J., Maier E., Mott R., McCarthy L., Grigoriev A., Schalkwyk L., Francis F. and Lehrach H. (1993) High resolution cosmid and P1 maps completely scanning the 14 Mbp genome of the fission yeast *Schizosaccharomyces pombe. Cell*
8. Mott R. Grigoriev A, Maier E, Hoheisel J and LehrachH (1993) Algorithms and software tools for ordering clone libraries: application to the mapping of the genome of Schizosaccharomyces pombe. Nucl. Acids Res. 21: 1965-1974

DATA MANAGEMENT TOOLS FOR SCIENTIFIC APPLICATIONS: FRAMEWORK, STATUS, AND GENOMIC APPLICATIONS *

Victor M. Markowitz

Computer Science Research and Development
Department Information and Computing Sciences Division
Lawrence Berkeley Laboratory
Berkeley, CA 94720
Email address: VMMarkowitz@lbl.gov

ABSTRACT

Data management systems for scientific (e.g., genomic) applications provide scientists with facilities for maintaining, manipulating and analyzing data. These systems usually employ capabilities provided by database management systems (DBMSs) for storing, accessing, and manipulating data, and often need facilities for handling application-specific constructs and operations that are not supported by DBMSs. The mismatch between the application and DBMS constructs and operations is the main cause for the difficulty of developing and maintaining data management systems for scientific applications, as well as adapting these systems to structural changes entailed by the evolution of the underlying applications.

Data management systems for different scientific applications are often based on different DBMSs and reflect different views of a shared universe of discourse. In order to exchange and share data, such heterogeneous data management systems need a unified (integrated) view of their underlying applications. The semantic (schema) integration process required for constructing such a view involves complex structural and data conversion mappings.

We discuss in this paper problems concerning the development, maintenance, evolution, and schema integration of data management systems for scientific applications. We propose insulating such applications from their underlying DBMSs through an intermediary object-based level. This approach allows an evolutionary development of data management systems and facilitates their transfer to other DBMSs. We illustrate our discussion with genomic application examples. The data management tools are aimed at increasing the productivity, efficiency, and accuracy of developing, maintaining, modifying, and semantically integrating data management systems for scientific applications. We outline in this paper our approach to the development of data management tools, and overview the current status of the data management tools developed at Lawrence Berkeley Laboratory.

---

*Issued as technical report LBL-32382. This work was supported by the Office of Health and Environmental Research Program and the Applied Mathematical Sciences Research Program, of the Office of Energy Research, U.S. Department of Energy, under Contract DE-AC03-76SF00098.

1. INTRODUCTION

In the past, database management systems (DBMSs) were rarely used in scientific applications. Typically, scientists dealt directly with files structured specifically for their applications. More recently, it has been recognized that the complexity and the amount of data for scientific applications require data management facilities for keeping track of the data being generated by experiments, simulations, and measurements. In addition, the data needs to be presented, browsed and queried in a way that scientists would understand. Scientific projects also need to interface fairly complex application programs (e.g. for data analysis, visualization, simulation) to the data management system that maintains the data.

Because of the special requirements of scientific applications (such as temporal, spatial, and sequence data), it is necessary to use data management technology that can be extended to accommodate new data structures and operators. There are several new methodologies currently being pursued for the data management of complex applications. They include object-oriented DBMSs, extended relational DBMSs (which allow new constructs to be added to the basic constructs provided by relational DBMSs), and extensible database systems (where selected building blocks are compiled into specialized database systems). These technologies promise to have an important role in supporting scientific applications. However, in our interactions with various scientific (e.g. genomic) applications we have found that in practice scientific users are still struggling with ways to capture conventional aspects of data (e.g. information about experiments, keeping track of laboratory equipment and experiments, references to articles), and that they prefer to use proven commercial DBMS technology, which means, of course, relational DBMSs.

Developing data management systems for scientific applications requires languages for specifying the structure of data and for querying data, mechanisms for efficiently accessing data, facilities allowing concurrent database access by multiple users, and mechanisms for enforcing the validity and integrity of data. Such capabilities are usually provided by DBMSs. The languages provided by DBMSs for describing data structures, however, are at a general-purpose, lower level, of abstraction than that of most scientific applications. Relational DBMSs, for example, do not provide constructs for directly representing genomic maps or molecular biology laboratory experiments. DBMS specifications involve (mainly technical) terms that are foreign to scientists, contain elements that have no direct counterpart in the application, and therefore obscure the semantics of the application. Moreover, DBMSs do not provide mechanisms for representing the data in a comprehensible form for scientists. Thus, while a *map* may look like in Figure 1, it can be represented in a relational database only by disconnected tuples that are scattered among multiple tables. The mismatch between the application and DBMS structures and operations is resolved by appropriate (structure and operation) mappings. The complexity of these mappings makes the development, maintenance, and modification of data management systems for scientific applications tedious, error- prone, and time-consuming processes.

The past years witnessed a tremendous proliferation of data management systems that have been developed for closely related scientific applications, and that are characterized by various degrees of heterogeneity. For example, there are currently several major data management systems containing data pertaining to the Human Genome project. These systems provide a variety of information including nucleic acid sequences (e.g. *GenBank, EMBL, DDBJ*, and *GenInfo*), peptide sequences (e.g. *PIR, SwissProt, GenPept*), as well as mapping data for various organisms (e.g. *GDB* and *CEPH* for Humans, *GBase* for Mouse, the Yale database for E.coli, and the UCB database for Yeast). These systems are heterogeneous in nature: they reside on different hardware, under different operating systems; they are based on different DBMSs; and most important, they reflect different views of the same scientific universe of discourse. Exchanging and sharing data across heterogeneous data management

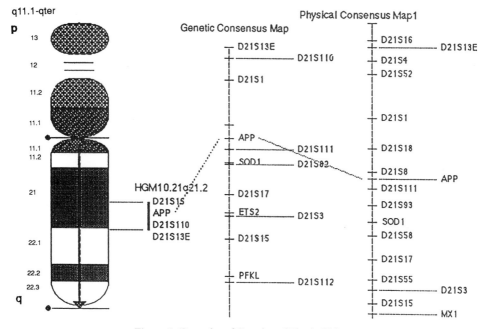

Figure 1. Examples of Genetic and Physical Maps.

systems requires constructing a unified (integrated) view of the applications underlying these systems. The *semantic schema integration* process required for constructing such a view involves complex structural and data conversion mappings.

In this paper we discuss problems concerning the development, maintenance, evolution, and integration of data management systems for scientific applications. We propose insulating such applications from the underlying DBMS through an intermediary object-based level. This approach allows an evolutionary development of data management systems using existing commercial DBMSs, and facilitates their transfer to other DBMSs. We illustrate our discussion with examples taken from genomic applications.

The data management tools are aimed at increasing the productivity, efficiency, and accuracy of developing, maintaining, modifying, and semantically integrating data management systems for scientific applications. Only tools for developing database structure (schema) definitions are commercially available. These tools (e.g. *ERwin,* IDE's *Software Through Pictures,* Chen's *ER-Modeler*) have been developed in order to facilitate the database definition process, and are based on variations of the *Entity-Relationship* (ER) model [1]. These tools achieved their goals only partially because the ER model versions underlying these tools provided only a limited set of constructs, and because the DBMS definitions generated by these tools were incomplete by not including referential integrity constraints implied by the semantics of ER schemas. We have researched various theoretical and practical aspects regarding the development of data management tools (see [2, 3, 4, 5]). Because of the limitations of commercially available tools, we started in 1988 to develop our own schema specification tools. Subsequently, we proceeded to research and develop additional tools for facilitating various aspects of the data management development process. In this paper we discuss our approach to the development of data management tools and overview the current status of the tools developed at Lawrence Berkeley Laboratory.

The rest of the paper is organized as follows. In section 2, we discuss the architecture of data management systems for scientific applications and the problems involved in developing, modifying, and integrating such systems. In section 3, we propose a three-level architecture for data management systems for scientific applications, and briefly review the data model and query language of the intermediary object level in this architecture. The role of data management tools in developing, modifying, and integrating data management systems is examined in section 4. In section 5 we overview the current status of data management tool development at Lawrence Berkeley Laboratory. The paper concludes with a summary.

2. DATA MANAGEMENT SYSTEMS FOR SCIENTIFIC APPLICATIONS

In this section we briefly overview the basic aspects of developing, modifying, and integrating data management systems for scientific applications.

Scientific applications are often quite complex in terms of the number of different types of objects and their relationships. Typically, scientific applications can be described in terms of objects characterized by (having) attributes; objects sharing common attributes are considered to have the same type, and are classified in object classes (homogeneous sets of objects). For example, genomic maps can be described as objects that share common attributes, such as *Name* and *Type*, and therefore can be classified in a class of objects called *Map*, as shown in Figure 2. Attributes can be (i) *atomic* (e.g. *Type, Name, Title*), or composite, that is, consisting of aggregations of atomic attributes (e.g. *(Locus, Location)*; (ii) *single-valued* (e.g. *Title*), or *set-valued* (e.g. *{Citation}, {(Locus, Location)}*); (iii) *local* (e.g. *Name, Title*), or *referential*, that is, attributes that can be used as references to other objects (e.g. attribute *Locus* of *Map* is a reference to *Locus* objects). Object classes and attributes are usually associated with special operators, such as *map alignment* and *map construction* for genomic map objects.

2.1 The Development of Data Management Systems for Scientific Applications

As mentioned in the introduction, data management systems for scientific applications usually employ a DBMS for storing and manipulating data generated by applications. DBMS constructs and operations, however, are at a lower level of abstraction than that of scientific applications. The mismatch between the application-specific and DBMS constructs and operations entails a two-level architecture for data management systems of scientific applications, as shown in Figure 3:

1. the *application* level contains the definition of application-specific constructs and operations;
2. the *database* level contains the application-generated data structured according to a database definition (schema) using constructs supported by the underlying DBMS; the underlying DBMS provides mechanisms for manipulating data in the database.

The mismatch between the application-specific and DBMS constructs and operations is resolved by a *schema mapping* between the application-specific and database constructs (schemas), an *operation mapping* between the application-specific and database operations, and a *data conversion* of the data structured according to the database definition into data structured according to the application-specific schema.

Map	Locus	Citation
Name	Name	Title
Type	Type	Author
{ (Locus , Location) }	{ Citation }	. . .
{ Citation }	. . .	
. . .		

Figure 2. Examples of Genomic Map Object Classes and Attributes.

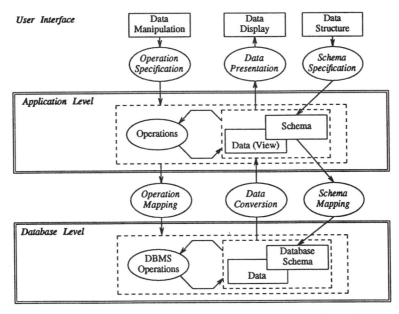

Figure 3. Basic Architecture for Data Management Systems for Scientific Applications.

Users interact with the data management system via a *user interface* that provides a *schema specification* facility for defining the application-specific data structure, a *data presentation* facility for displaying application data in a suitable (e.g. graphical) format for users, and an *operation specification* facility for specifying application-specific operations. The development of data management systems for scientific applications involves defining the application- specific constructs and operations, defining the underlying database using DBMS constructs and operations, implementing the mappings between the application-specific and DBMS constructs and operations, and constructing a user interface.

2.2 The Evolution of Data Management Systems for Scientific Applications

Scientific applications evolve over time, and consequently their data management systems need to be modified by, for example, introducing new classes of objects, changing the structure of existing types of objects, etc. The evolution of an application affects both the application and database levels of the associated data management system, as illustrated in Figure 4:

1. the evolution of the application entails restructuring the application data structure (schema);
2. the effect of restructuring the application schema is defined by restructuring the database schema, and reorganizing the data in the database according to the restructured database schema;
3. the schema and operation mappings, and the data conversion between the application and database levels must be modified in order to accommodate the restructured application and database schemas and the reorganized data content of the database.

2.3 Semantic Integration of Data Management Systems for Scientific Applications

Scientific applications usually need to share and exchange information. For example, molecular biology mapping applications usually need information from various sources (e.g. *GenBank, GDB, EMBL)*. DBMSs provide shared access to data created by multiple

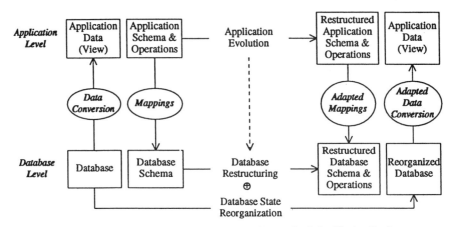

Figure 4: The Evolution of a Data Management System for Scientific Applications.

autonomous applications, but employing a DBMS entails organizing the data in a central database without allowing the applications to have their own data organization. Thus, the database provides a centralized and homogeneous view of data to applications.

Although data management systems of different applications are accessible via high speed computer networks, they often reside on different hardware, are developed under different operating systems and with different DBMSs, and have different user interfaces. They also reflect very different views of the common scientific universe of discourse. Such heterogeneous data management systems can be organized in a *federation* [6] in which each participating system is autonomous, that is, each system (i) preserves its own application view, (ii) can decide with what other systems it communicates and what information it exchanges; and (iii) can decide how and when to execute requests from other systems.

One of the main problems of organizing a federation of heterogeneous data management systems is to construct a unified federated view (schema) of the applications underlying the component data management systems. The process of constructing such a view is called *semantic (schema) integration*. In a data management system federation (see Figure 5), every (local) data management system has its own application schema; local application schemas are semantically integrated into a *federated* application schema (view) that can be used for external access of the data management systems in the federation; local data management systems can have their own view of the federated schema, described in the language (data model) of the local system.

The semantic integration process involves four main stages (activities) [7]:

1. *Preintegration* refers to the preliminary stage of integration, in which a common data model is selected and the local application schemas are described using this common data model.
2. The *comparison* of the schemas involved in integration is carried out in order to detect conflicts and inconsistencies regarding names, structures, attribute domains, etc.
3. The *alignment* of the schemas involved in the integration resolves the conflicts detected in the previous stage, in order to achieve pairwise compatibility. Subsequently, the similarity of object classes is established in preparation for merging the schemas.
4. *Merging* of compatible schemas into an integrated federated schema, followed by a restructuring of the integrated schema in order to eliminate redundancies.

Various schema integration methodologies differ in the way they implement these stages (see [7]).

30

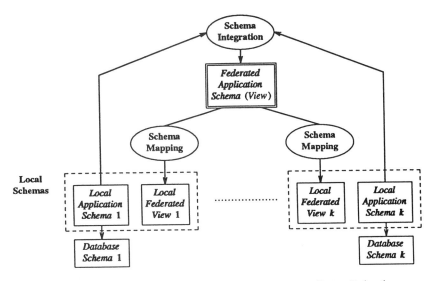

Figure 5. Application Schemas (Views) in a Data Management System Federation.

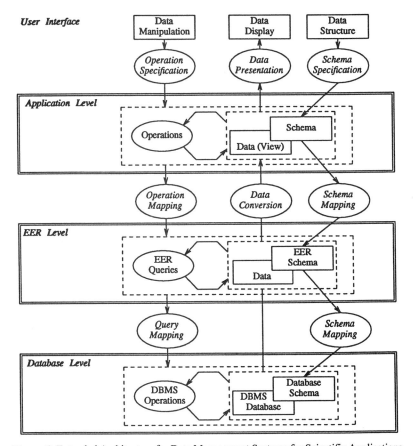

Figure 6. Extended Architecture for Data Management Systems for Scientific Applications.

3. EXTENDED ARCHITECTURE FOR DATA MANAGEMENT SYSTEMS

As discussed in section 2, data management systems for scientific applications involve mappings between application- specific and DBMS constructs and operations. The specification of these mappings can be significantly simplified by introducing an intermediate level between the application and database levels, as shown in Figure 6. We propose using the *Extended Entity-Relationship* (EER) model for this intermediate level. The EER level in the architecture of a data management system for scientific applications allows decomposing the operation and schema mappings between the application and database levels into simpler mappings between the application and EER levels, and the EER and database levels, respectively.

The EER model is an extended version of the popular Entity-Relationship (ER) model of [1]. EER schemas consist of specifications for atomic types of objects called *entity-sets* (e.g. MAP, PROBE, LOCUS, CITATION), and associations of entity-sets called *relationship-sets* (e.g. CONSISTS OF, WRITTEN BY). Individual entity-set and relationship-set instances (i.e. individual entities and relationships) are qualified by *attributes* (e.g. NAME, TITLE). Entity-sets, relationship-sets, and attributes can be represented diagrammatically in an EER schema by rectangle, diamond, and ellipse shaped vertices, as illustrated in Figure 7; relationship-sets are connected by arcs to the object-sets they associate, and entity-sets and relationship-sets are connected by arcs to their attributes. *Generalization* is an abstraction mechanism that allows viewing a set of *specialization* entity-sets (e.g. PROBE, LOCUS) as a single *generic* entity-set (e.g. MAP OBJECT). The attributes and relationships which are common to the entity-sets that are generalized (e.g. NAME) are then associated only with the generic entity-set. A specialization entity-set (e.g. PROBE) inherits the attributes and relationship-sets of all its generic entity-sets (e.g. PROBE inherits attribute NAME as well as relationship-set REFERENCED IN from REFERENCED OBJECT). Generalization is represented diagrammatically in an EER schema by arcs labeled *ISA*, connecting specialization entity-sets to generic entity-sets. A more detailed definition of the EER constructs is provided in [5]. We refer below commonly to entity-sets and relationship-sets as object-sets.

Several query languages have been proposed in the literature for the ER and EER models (e.g. [8, 9, 10, 11]). Among these languages the Concise Object Query Language (COQL) proposed in [10] is unique in its support for attribute and relationship inheritance, conciseness, and in allowing attributes associated with (*auxiliary*) object-sets to be viewed as attributes of other (*primary*) object-sets. As will be shown in section 4, the latter is essential for specifying application-specific object structures. The only assumptions made in COQL are that the underlying data model is capable of describing objects, their attributes, and connections between objects. The nature of these connections is not visible in COQL, so that users do not have to remember the schema structure. COQL queries are concise and quite close to their English formulation.

COQL supports both attribute and relationship *inheritance*. Thus, for objects (e.g. MAP) that are specializations of generic objects (e.g. REFERENCED OBJECT), the attributes (e.g.

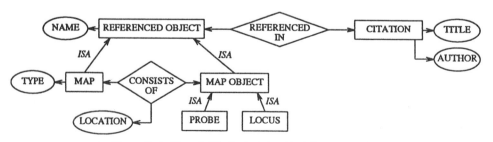

Figure 7. An Extended Entity-Relationship Schema Example.

NAME) of the generic objects will be treated as if they are directly associated with (i.e are inherited by) the specialization objects. This capability simplifies the expression of COQL queries and improves their conciseness.

A COQL query consists of an **OUTPUT** statement, a **CONDITIONS** statement, and a **CONNECTIONS** statement; the **CONDITIONS** and **CONNECTIONS** statements are optional. The order of these statements is arbitrary, thus allowing users to construct queries incrementally according to their thought process. The **CONDITIONS** statement consists of conditions associated with individual object-sets. A condition referring to an object-set can involve its local or inherited attributes, and consists of and-or compositions of atomic comparisons of the form $A \theta val$ or $A \theta B$, where A and B are local or inherited attributes, or aggregate functions (e.g. **COUNT, SUM**), θ is a comparison operator (such as `=', `>', **MATCH**), and val is a list of atomic values. The **OUTPUT** statement contains the list of primary object-sets targeted for selection. A primary object-set can be associated in an output statement with its local and inherited attributes, as well as with attributes of other (auxiliary) object-sets. Consider, for example, the EER schema of Figure 7. The following COQL query retrieves the Gingrich or Tanzi maps together with their *loci*, and *citations*:

> **CONDITIONS**   MAP: NAME = "Gingrich", "Tanzi";
> **OUTPUT**   MAP: NAME, NAME **OF** LOCUS, TITLE **OF** CITATION;
> **END**

In the COQL query above, MAP is a primary object-set, while LOCUS and CITATION are auxiliary object-sets, so that the (set of) NAME of LOCUS and TITLE of CITATION are viewed as (multi-valued) attributes of MAP.

The **CONNECTIONS** statement specifies the connections between the object-sets appearing in a COQL query. If the connections between the objects appearing in the query are unambiguous, then they do not need to be specified. The ability to express queries without explicit connections contributes to their conciseness. In the COQL query above, for example, explicit connections are not necessary because it is clear that MAP is related to LOCUS and CITATION via CONSISTS OF and REFERENCED IN, respectively. When the connections are ambiguous, several strategies can be used for selecting the intended connections or the user can include explicitly the intended connections into the query.

4. THE EFFECT OF THE INTERMEDIATE EXTENDED ENTITY-RELATIONSHIP LEVEL ON DEVELOPING, MODIFYING AND INTEGRATING DATA MANAGEMENT SYSTEMS

In this section we discuss the effect of the extended architecture presented in section 3, involving an intermediate EER-based level, on the development, evolution, and schema integration of data management systems.

4.1 Developing Data Management Systems with Intermediate EER Levels

The three-level architecture described in section 3 allows decomposing the mappings between the application and database levels into simpler mappings between the application and EER levels, and the EER and database levels, respectively. This decomposition has several advantages that are briefly discussed below.

EER schemas and queries are specified in terms of objects and object connections, and therefore are inherently more concise, simpler to specify, and simpler to comprehend than DBMS schemas and queries. Accordingly, expressing scientific application-specific data structures and operations using EER constructs and queries is significantly simpler than using directly DBMS constructs and queries. Scientists can learn to understand and critique a database design expressed in EER terms, whereas detailed DBMS specifications, which are obscured with low level technical details, are much more difficult to understand. Moreover, developing EER schemas and queries is independent of a specific DBMS, and therefore can eventually be transferred to other DBMSs.

The object and attribute concepts of the EER model are closer to the concepts naturally employed by scientists to describe the structure of their applications. However, the EER object and attribute concepts are still too restricted for directly describing application-specific (e.g. genomic map) objects and attributes. For example, the EER model does not directly support set-valued and composite attributes. As discussed in [5], such restrictions are essential for an accurate mapping of EER schemas into DBMS schemas, and they can be overcome by using *auxiliary* EER objects. Auxiliary EER objects do not represent application-specific objects, and are used for specifying structures that are not directly supported in the EER model. For example, consider class of objects *Map* shown in Figure 2. *Map* can be represented by an EER entity-set, MAP, as shown in the EER schema of Figure 7. However, in order to represent that a *Map* object involves sets of *Locus* objects (where each *Locus* is placed at a certain *Location*), as well as *Probe* objects, an auxiliary relationship-set, CONSISTS OF, and an auxiliary generic entity-set, MAP OBJECT, are needed, where MAP OBJECT generalizes the entity-sets representing *Probe* and *Locus*. Similarly, in order to represent that *Map, Probe*, and *Locus* objects are associated with *Citation* objects, an auxiliary generic entity-set, REFERENCED OBJECT, and an auxiliary relationship-set, REFERENCED IN, are needed, where REFERENCED OBJECT generalizes the entity-sets representing *Map, Probe* and *Locus*.

Using multiple (including auxiliary) EER object-sets for representing a single application-specific class of objects implies that data regarding an (instance of an) application-specific object is usually scattered among multiple EER object-sets. Consequently, the EER representation of an application-specific structure must be complemented with operational definitions specifying the way application-specific objects are retrieved and updated via instances of EER object-sets. Such definitions can be expressed using COQL. For example, consider class of objects *Map* shown in Figure 2 and represented as shown in Figure 7; the EER structural representation of *Map* can be complemented with the following COQL query for retrieving *Map* objects:

```
OUTPUT MAP:      NAME, TYPE, NAME OF LOCUS, LOCATION OF CONSISTS_OF,
                 (TITLE, AUTHOR) OF CITATION
CONNECTIONS      MAP CONSISTS_OF LOCUS;
                 MAP REFERENCED_IN CITATION;
END
```

This query associates every MAP entity with the value of its (local) attribute TYPE, the value of the inherited attribute NAME, with sets of pairs of LOCUS NAME and LOCATION values, and with sets of CITATION TITLE values. Note that selections of *Map* objects that satisfy certain conditions require executing a modified version of the COQL query above by adding only a condition statement, such as:

```
CONDITIONS       MAP: NAME MATCH "%Cytogenetic%";
```

EER schemas and queries can be readily implemented over commercial relational DBMSs. An EER schema can be represented by a relational DBMS schema consisting of definitions for relation-schemes (tables), keys (indexes), referential integrity and null constraints [5]. Informally, relation-schemes represent EER object-sets, referential integrity constraints represent the existence dependencies inherent to object-set connections, and keys represent EER object identifiers and cardinalities. EER (e.g. COQL) queries can be mapped into queries expressed in DBMS query languages such as SQL. DBMS schema definitions and queries, however, are very large. For the Sybase DBMS, for example, a database representing tens of EER objects requires thousand of lines of code for defining tables, indexes, and procedures for maintaining the integrity of data. Developing and maintaining large database definitions, procedures, and queries is a tedious, error prone, and time consuming process. Consequently, tools for carrying out automatically the EER to DBMS schema and query mapping are essential for fast and accurate translations from EER to DBMS specifications. The development of such tools is discussed in section 5.

4.2 Modifying Data Management Systems with Intermediate EER Levels

The evolution of a scientific application entails modifying the associated data management system. As discussed in section 2, such a modification affects both the application and database levels of the data management system (see Figure 4). A simple restructuring of the application schema generally entails complex database schema restructuring and database reorganization mappings, as well as complex modifications of the mappings and conversions between the application and database levels.

The three-level architecture described in section 3 allows considering the database restructuring and reorganization mappings as compositions of simpler mappings defining (i) the effect of restructuring an application schema on the EER schema, and (ii) the effect of restructuring the EER schema on the underlying database. These mappings are easier to develop and implement. Consider for example, the object classes that are shown in Figure 2 and that are represented by the EER schema of Figure 7. Suppose that attribute *Author* of object class *Citation* is converted from a single-valued to a multi-valued attribute. This trivial application schema modification entails removing attribute AUTHOR associated with CITATION, and adding two new object- sets, WRITTEN BY and AUTHOR, to the EER schema of Figure 7, as shown in Figure 8. Additionally, COQL queries, such as the query shown in section 4.1 above, must be adapted to the schema change. For example, the COQL query shown in section 4.1 must be changed into:

OUTPUT MAP: NAME, TYPE, NAME **OF** LOCUS,LOCATION **OF**
CONSISTS_OF, TITLE **OF** CITATION,NAME **OF** AUTHOR;
CONNECTIONS MAP CONSISTS_OF LOCUS;
MAP REFERENCED_IN CITATION;
CITATION_ WRITTEN_BY AUTHOR;
END

The effect of restructuring an EER schema on an underlying relational database has been explored in [12]. In general, one elementary EER schema restructuring operation (e.g. adding or removing an attribute, adding or removing an EER object-set) requires several complex database schema restructuring and database reorganization operations. For example, consider the EER schema modification above and suppose that the underlying database is developed with the Sybase DBMS. Removing an EER attribute does not have an analogous Sybase (or any other DBMS) restructuring operation, therefore removing attribute AUTHOR associated with CITATION requires:

(i) saving the key definitions, index definitions, and referential integrity (*trigger*) procedures associated with and the data content of the relation (table) representing CITATION;
(ii) removing from the database the relation representing CITATION, replacing it by (a) adding a redefined relation, without the attribute (column) representing AUTHOR, and (b) reloading the data without the AUTHOR values into this redefined relation;
(iii) reloading the key definitions, index definitions, and referential integrity procedures associated with the new relation representing CITATION.

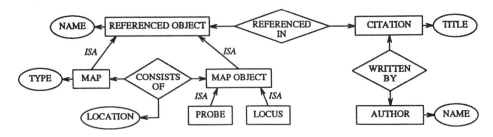

Figure 8. Modified Extended Entity-Relationship Schema.

Additionally, the procedures implementing COQL queries such as that above, must be modified according to the schema restructuring.

Even trivial application schema modifications that entail trivial EER schema modifications may require complex database schema restructuring and reorganization operations. For example, consider an attribute in an application schema that is represented by an EER attribute, which in turn is represented by an attribute in a database relation. Changing the domain of the attribute in the application schema entails changing the domain (values-set) of the corresponding EER attribute. However, changing the domain of a relational attribute is not directly supported by commercial DBMSs, therefore such a change requires removing the relation involving the affected attribute and replacing it with a new relation in which the attribute is associated with a new domain, in a way similar to that described above for removing an attribute. Moreover, if the affected attribute is a key attribute then the relation replacement must be repeated for every relation that involves a (foreign-key) attribute referencing the affected attribute.

As the discussion above shows, manually modifying data management systems (especially their database components) is a very complex, error prone, and time consuming process. Accordingly, tools for assisting users in modifying data management systems are very desirable. However, there are no reported attempts to develop such tools.

4.3 Schema Integration for Data Management Systems with Intermediate EER Levels

In section 2, we have listed the main stages involved in the process of (semantic) schema integration. Schema integration methodologies differ in the way they implement these stages. A comprehensive comparison of twelve schema integration methodologies is provided by [7]. Most of these methodologies are based on versions of the EER model. Consequently, the intermediate EER level of the extended architecture presented in section 3 allows using existing results in the area of schema integration.

In the *preintegration stage* a version of the EER model is often used as the common data model [7] mainly because of the widespread employment of ER and EER methodologies for developing database schemas. Many existing databases, however, are not developed using an object model and therefore require a reverse mapping for deriving an EER schema (this process is commonly known as *reverse engineering*). Reverse mappings are complex because local (e.g. relational) data models are often poorer in their semantic expressiveness than object (e.g. EER) models. Reverse mapping techniques are presented in several papers, such as [13, 14, 15].

In order to keep track of name correspondences, in the *alignment* stage a data dictionary or thesaurus can be used. Resolving structural differences usually involves schema (restructuring) transformations (e.g. see [16, 17]).

A simple example of schema integration is illustrated in Figure 9. Suppose that bibliographic citations are modelled differently in two data management systems. In the *preintegration* stage the EER model is used as the common data model for representing these data, as shown in Figure 9(i). The *comparison* of the two schemas shown in Figure 9(i) reveals two conflicts:

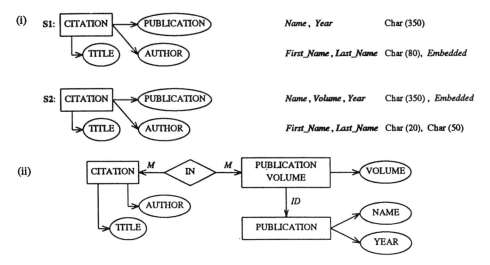

Figure 9. An Example of Schema Integration.

(i) A structural conflict for attribute PUBLICATION which is defined differently in these schemas, namely as *Name, Year* for schema S1, and *Name, Volume, Year* for schema S2, respectively. Accordingly, these schemas are *aligned*, that is, restructured in order to become compatible. The restructuring involves (a) converting attribute PUBLICATION from an attribute of CITATION into an entity-set associated by relationship-set IN with CITATION, and then (b) splitting the PUBLICATION entity-set into two entity-sets, PUBLICATION VOLUME and PUBLICATION.

(ii) A domain conflict for attribute AUTHOR, which has two different formats: a Char (80) format for the (embedded) *First Name* and *Last Name* in the first schema, and Char (20) and Char (50) formats for *First Name* and *Last Name*, respectively, in the second schema. In the alignment stage a new (virtual) attribute is specified, and associated with two domains of type Char (20) for *First Name* and Char (80) for *Last Name*, respectively. *Domain mappings* from the domains associated with the original attributes to the domain of the virtual attribute are also defined in this stage (for details on domain conflict resolution techniques see [18]).

Finally, the two (now compatible) schemas are merged into the integrated schema shown in Figure 9(ii). For this simple schema integration example no restructuring is needed in order to eliminate redundancies. If a common application-specific object model is available, then the semantic integration must be followed by a reverse mapping of the EER schema into an application specific schema.

There are only a few reported attempts to develop schema integration tools (e.g. see [19]). The goal of these tools was to demonstrate the feasibility of the underlying methodologies, rather than being complete implementations. Accordingly, these tools are only partial experimental implementations. Furthermore, these tools have not been employed in real projects requiring schema integrations.

5. DATA MANAGEMENT TOOLS DEVELOPED AT LBL: A STATUS REPORT

Commercial schema specification tools based on various versions of the *Entity-Relationship* (ER) model have been developed in order to facilitate defining database schemas. These tools have achieved their goals only partially because they support only a limited set of

ER constructs, and because the DBMS definitions generated by these tools do not include referential integrity constraints entailed by the semantics of ER schemas. Unlike schema specification tools, commercial tools for specifying database queries are not based on an object model, and assist users in specifying queries in the SQL dialect of a specific DBMS. Examples of such tools are Sybase's *Visual Query Language* and Ingres's *Query By Forms*. Because of the limitations of commercially available data management tools, we decided to develop our own schema definition and query tools based on the EER model defined in [5]. These tools are overviewed below. Various theoretical and practical aspects regarding the translation of EER schemas and queries into DBMS database definitions and queries are examined in [2, 3, 4, 5, 10]. We also briefly discuss below some application- specific data management tools whose development has started recently.

5.1 Schema Definition and Translation Tools

EER schemas can be specified and edited using either a textual or a graphical (EER diagrammatic) language. For graphical specification and editing of EER schemas we have developed a graphical tool called ERDRAW. ERDRAW supports graphical manipulations of EER schemas (e.g. adding, removing, moving, and changing the type of object-sets), editing functions (e.g. assigning names to object-sets, associating attributes with object-sets), allows partitioning large EER schemas into pages, and provides a browsing facility for objects and attributes. EER schemas edited with ERDRAW can be saved for later use, printed, or translated into a textual language specification. A typical ERDRAW session is shown in Figure 10: after loading an existing EER diagram for editing, the user browses the list of objects shown in the **ER Index** window; once an object is selected (e.g., LOCUS), the pages of the EER diagram where this object appears are displayed and one of the pages is selected; next, the user selects object LOCUS for editing, enters a description for the LOCUS object in the **Entity Set** window, and chooses the *Attributes* button to enter a description for the SYMBOL attribute of LOCUS; the **Attribute List** window displays the list of attributes for object LOCUS. ERDRAW is described in [20].

We have developed a Schema Definition and Translation (SDT) tool for mapping EER schemas into schemas of relational DBMSs. Our approach to mapping EER schemas into relational schemas, as discussed in [5], is to separate the mapping process into several independent stages, such as normalization, assigning names to relational attributes and merging relation-schemes. The modularity of this approach allows easy modifications and extensions of SDT. SDT works in two main stages. In the first stage, normalized DBMS-independent (abstract) relational schemas are generated from EER schemas; these schemas consist of definitions for relations, keys, and declarative referential integrity constraints. This stage includes the assignment of names to relational attributes, and merging relations. In the second stage, SDT generates schema definitions for specific DBMSs. For DBMSs that have procedural referential integrity mechanisms (e.g. triggers in Sybase 4.0, rules in Ingres 6.3), the main part of this stage consists of generating the appropriate insert, delete, and update procedures for maintaining the referential integrity constraints entailed by the EER schema. SDT supports several interchangeable name assignment and merging algorithms. In addition to the DBMS definition corresponding to an EER schema, SDT generates a metadatabase that contains information (metadata) on the EER schema, the generated relational schema, and the mapping between these two schemas. This metadatabase is essential for developing the query translator, as explained in the next subsection. SDT is described in [21].

SDT and ERDRAW are implemented on Sun workstations under Sun Unix OS 4.1, using C, LEX, and YACC. Currently, SDT can map automatically EER schemas into DBMS schema definitions for DB2, Sybase 4.0 - 4.9, Ingres 6.3, Informix 4.0 and Oracle 7 DBMSs.

SDT and ERDRAW are used by over 35 data management groups worldwide. At LBL, SDT and ERDRAW have been success-fully applied to the development of several data management systems, such as the *Chromosome Information System (CIS)* [22]. For developing *CIS*, for example, different types of genomic mapping objects, as well as their interrelation-ships, have been analyzed and described using the EER model [23]. Using SDT for automati-

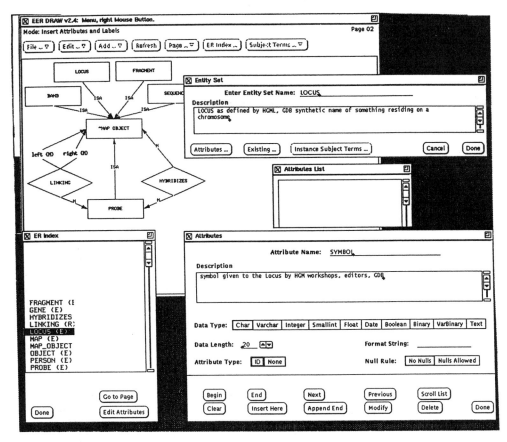

Figure 10. Major Types of Windows of the ERDRAW EER Schema Editor.

cally generating the Sybase definition underlying *CIS* increased the productivity and accuracy of the database development process. Thus, while the initial *CIS* EER schema was about 150 lines long, its complete Sybase schema definition was about 3500 lines long.

5.2 Query Specification and Translation Tools

The query specification and translation tools are based on the *Concise Object Query Language* (COQL) described in section 3, and assist users in interactively specifying queries in terms of objects and attributes. These tools do not require knowledge of DBMS or even EER concepts. Instead, users are only aware of the existence of objects and attributes, and specify queries which consist of objects, attributes, operators, and values.

COQL queries can be specified and edited using either a textual or a graphical language. For graphical specification of COQL queries we have developed a graphical tool. The purpose of this query editor is to guide the user through various stages of query specification. COQL queries are specified interactively by clicking on action buttons and selecting items from scroll lists in a window environment. The various query parts are incrementally constructed and shown on the screen. We have adopted a what-you-see-is-what-you-get approach for composing queries on the screen. Thus, the order of elements in a query constructed with the COQL query editor reflects the natural order of specifying the query. A typical COQL query editing session is shown in Figure 11: after loading the definition of an existing EER schema using the metadatabase generated by SDT, the user browses through a list of objects shown in the **New Object** window; once an object is selected (e.g., MAP), the attributes associated with the object are listed in the **Attributes** window; attributes can be selected for inclusion in the output statement, and/or for specifying conditions using the **Condition** window; conditions can be specified directly or using a special **Conditions for Attribute** window for icon-based condition specification; a path of selected objects is incrementally constructed by the editor; if two adjacent objects in the path are not directly connected in the EER schema, the editor finds the connection between the objects (if it is unique) or lists all the connections and prompts the user to select the desired connection. Once the COQL query is constructed, the query editor can be used for immediately executing the query (the **Retrieve Data** button) or for storing the query as a stored procedure.

Concise object queries are not required to include all the connections between the objects. The connections that are not specified in a COQL query can be inferred using the information stored in the metadatabase on the structure of the underlying EER schema. In the COQL query given in section 3, for example, the connections of MAP with LOCUS and CITATION are not specified and can be inferred from the information in the metadatabase. When there is more than one way to connect objects, ambiguity can be resolved by using special algorithms (e.g. finding the shortest path) and by consulting the user. The result of resolving inheritance and connectivity is a *complete object query*.

COQL queries are mapped into queries in the SQL dialect specific to the underlying relational DBMS by a COQL query translation tool. The COQL query translator works in two main stages. In the first stage, a COQL query is translated into an intermediate, DBMS-independent (abstract), query expression. In this stage all attribute and relationship-set inheritance problems are resolved and object conditions and connections are interpreted and translated into SQL-like conditions. Based on semantic information regarding the EER schema, the query expression is semantically optimized by reducing, whenever possible, the SQL (outer join) conditions to simpler (regular join) conditions. In the second stage, the COQL query translator maps the intermediate query expression into a query in the SQL dialect of the underlying relational DBMS. The mapping in this stage follows a plan for minimizing the number of generated SQL subqueries. This plan depends on the specific capabilities provided by the target DBMSs; thus, for DBMSs such as Sybase that impose certain restrictions on combining outer and regular joins, these restrictions affect the number of generated SQL subqueries.

Both the COQL query editor and translator use the metadatabase that is generated by SDT and that contains information on the EER schema, the underlying DBMS schema, and the correspondence between the EER and DBMS schemas. The information in the metadatabase is essential for inferring the connections (paths) between the objects specified in a COQL query, and for providing the mapping information between the EER schema and the corresponding DBMS schema necessary in the process of translating COQL queries into SQL queries.

The COQL query editor has been implemented on Sun workstations under Sun Unix OS 4.1, using C++ and Xdesigner. The COQL query translator targeting the Sybase DBMS has been implemented on Sun workstations under Sun Unix OS 4.1, using C++, LEX, and YACC.

The COQL query editor and translator are currently used for developing several genomic databases, such as the *Integrated Genomic Database* at the German Cancer Research Centre, University of Heidelberg, as part of the European Data Resource for Human Genome Analysis Project [24].

5.3 Application-Specific Data Management Tools

We briefly describe below data management tools that we are currently developing in order to support application-specific structures and queries. The main purpose of developing these tools is to support the development of data management systems for Laboratory Information Management Systems. These tools, however, benefit other scientific applications as well.

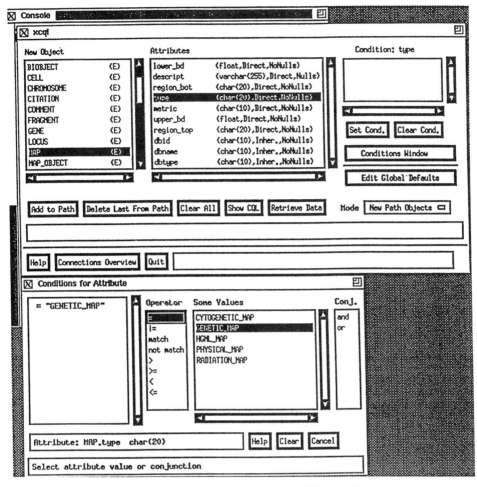

Figure 11. Major Types of Windows of the COQL Query Editor.

As already mentioned in section 4, the EER object and attribute constructs are too restricted for directly describing application-specific objects such as genomic maps (see section 4) or laboratory protocols. The latter is an essential part of *Laboratory Information Management System* (LIMS) applications. Protocols can be clearly and accurately described in terms of processes and process interactions. Given an input, a protocol instance (e.g. an elementary experiment) results in an output, where both the input and output consist of objects. Protocols often involve a series of steps, which are also protocols. The recursive specification of protocols in terms of component sub-protocols is called protocol expansion. Protocol expansion reveals the composition of component sub-protocols and/or alternative ways of performing the protocol.

Consider the protocol for labeling electrophoretic gels described in Figure 12 using a diagrammatic notation similar to the data flow (DFD) notation. In labeling *electrophoretic gels*, one can either label the gel directly by *staining*, or one can transfer the DNA from the gel to a *filter* via *southern blotting* and then *probe* the filter with a radioactively labeled *hybridization* probe. The generic labeling protocol is represented in Figure 12(i), where a LABELED SEPARATION sample (output) is the result of a LABEL experiment applied to an ELECTROPHORETIC GEL. The protocol shown in Figure 12(ii) is the result of expanding

41

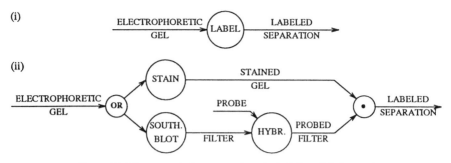

(i)

ELECTROPHORETIC GEL → (LABEL) → LABELED SEPARATION

(ii)

ELECTROPHORETIC GEL → (OR) → (STAIN) → STAINED GEL → (•) → LABELED SEPARATION

(SOUTH. BLOT) → FILTER → (HYBR.) PROBE PROBED FILTER

Figure 12. Diagrammatic Representation of a *Labeling Gels* Protocol.

the generic LABEL protocol of Figure 12(i) into two alternative protocols, STAIN, respectively SOUTHERN BLOT followed by HYBRIDIZATION. Note that experiments interact through the objects passing between them, and an object may be the input or output of different experiments (e.g. an ELECTROPHORETIC GEL can be the input of either STAIN or SOUTHERN BLOT).

For describing complex structures specific to scientific applications such as genomic mapping and LIMS applications, we have developed an Object-Protocol Model (OPM) [25]. OPM supports the specification of object and protocol classes. Connections of object and protocol classes are expressed via attributes and attribute value classes. Protocol classes can be expanded into alternative or sequences of (sub) protocol classes. For example, the protocols of Figure 12 are (partially) described using OPM as follows:

PROTOCOL CLASS	LABEL
EXPANSION	STAIN *or* (SOUTHERN_BLOT, HYBRIDIZATION)
ATTRIBUTE	Electrophoretic_Gel: ELECTROPHORETIC_GEL input
ATTRIBUTE	Labeled_Separation: LABELED_SEPARATION output
PROTOCOL CLASS	STAIN
ATTRIBUTE	Electrophoretic_Gel: ELECTROPHORETIC_GEL input
	isa LABEL Electrophoretic Gel
ATTRIBUTE	Stained_Gel: STAINED_GEL output
	isa LABEL.Labeled_Separation

We have developed a graphical tool for specifying, editing, and browsing OPM (object-protocol) schemas. This tool allows users to examine object and protocol structures, select individual protocols for expansion or for displaying their parameters. The protocol specification is hierarchical, so that the protocol designer can progressively specify the protocol in increasing levels of detail. We have developed a tool for mapping OPM schemas into EER schemas. This tool is coupled with existing EER to DBMS mapping tools for generating the DBMS schema definitions. Currently, we develop a query language and a query editor for OPM. The query editor will allow users to browse through object and protocol definitions, and specify queries in terms of objects and protocols; queries will be subsequently translated into COQL queries, and finally into SQL queries using the COQL translator.

6. SUMMARY

We have examined in this paper problems regarding the development, evolution, and schema integration of data management systems for scientific applications in general, and genomic applications in particular. We have argued that fast and accurate development, evolution, and integration requires special data management tools providing the capability of carrying out automatically various mappings involved in these processes. We have described a three-level architecture for data management systems for scientific applications that allows an incremental development of such tools, and overviewed the current status of the tools developed at Lawrence Berkeley Laboratory.

We plan to further extend our existing tools, as well as develop new data management tools. Thus, we plan to extend SDT and ERDRAW with additional capabilities that, for example, will allow defining abstract domain constraints such as, allowed ranges; such constraints can be then translated into appropriate integrity procedures specified using DBMS specific mechanisms (e.g. the rule mechanism in Sybase, the integrity mechanism in Ingres). We plan to extend the COQL query translator and to enhance the capabilities of the COQL query editor by extending COQL and by developing a data browsing capability.

ACKNOWLEDGEMENTS

The author wants to thank Arie Shoshani for his support and contribution to the development of the data management tools, and to express his appreciation to Ernest Szeto for his suggestions during the development of COQL and for his excellent implementation of the COQL to SQL translator. The author also wants to thank Suzanna Lewis, John McCarthy, Frank Olken, and Manfred Zorn of the LBL Human Genome Computing Group, Penny Bagett, Candy Robinson, and Bill Marshall of the SSCL Magnet Test Division, and Otto Ritter of the German Cancer Research Centre for sharing their applications and for using and suggesting improvements of the data management tools developed at LBL.

REFERENCES

1. P.P. Chen, "The Entity-Relationship Model-Towards a Unified View of Data", ACM Trans. on Database Systems 1,1 (March 1976), pp. 9-36.
2. V.M. Markowitz, "Problems Underlying the Use of Referential Integrity in Relational Database Management Systems", Proc. of 7th Int. Conf. on Data Engineering, IEEE Computer Society Press, April 1991.
3. V.M. Markowitz, "Safe Referential Integrity Structures in Relational Databases", in Proc. of the 17th Int. Conf. on Very Large Data Bases, September 1991.
4. V.M. Markowitz, "Merging Relations in Relational Databases", Proc. of the 8th Int. Conf. on Data Engineering, IEEE Computer Society Press, February 1992.
5. V.M. Markowitz and A. Shoshani, "Representing Extended Entity-Relationship Structures in Relational Databases: A Modular Approach", ACM Trans. on Database Systems, 17, 3 (Sep. 1992).
6. D. Heimbigner and D. McLeod, "A Federated Architecture for Information Management", ACM Trans. on Office Information Systems 3, 3 (July 1985), pp. 253-278.
7. C. Batini, M. Lenzerini, and S. Navathe, "A Comparative Analysis of Methodologies for Database Schema Integration", ACM Computing Surveys 18,4 (Dec. 1986), pp. 323-364.
8. B. Czejdo, R. Elmasri, D.W. Embley, and M. Rusinkiewicz, "A Graphical Data Manipulation Language for an Extended Entity-Relationship Model", IEEE Computer 23, 3 (March 1990), pp. 26-36.
9. V.M. Markowitz and Y. Raz, "ERROL: An Entity- Relationship Role Oriented Query Language", Entity- Relationship Approach to Software Engineering, Davis,G.C. and al (eds), North-Holland, 1983, pp. 329-345.
10. V.M. Markowitz and A. Shoshani, "Object Queries over Relational Databases: Language, Implementation, and Applications", Proc of the 9th Int. Conf. on Data Engineering, IEEE Computer Society Press, April 1993.
11. A. Shoshani, "CABLE: A Language Based on the EntityRelationship Model", Lawrence Berkeley Laboratory Technical Report LBL-22033, 1978.
12. V.M. Markowitz, "The Effect of Restructuring Extended Entity-Relationship Schemas on Sybase Database Definitions", TR LBL-31319, Lawrence Berkeley Laboratory, October 1991.

13. M.A. Casanova and J.E. Amaral de Sa, "Mapping Uninterpreted Schemes into Entity-Relationship Diagrams: Two Applications to Conceptual Schema Design", IBM Journal of Research and Development, vol. 28, no. 1, pp. 82-94, January 1984.

14. K.H. Davis and A.K. Arora, "Converting a Relational Database Model into an Entity-Relationship Model", in Proc. of the Sixth International Conference on Entity- Relationship Approach, S. March (ed), Elsevier Science Publishers B.V., pp. 271-285, 1987.

15. V.M. Markowitz and J.A. Makowsky, "Identifying Extended Entity-Relationship Object Structures in Relational Schemas", IEEE Trans. on Software Engineering, 16, 8 (Aug. 1990), pp. 777-790.

16. V.M. Markowitz and J.A. Makowsky, "Incremental Restructuring of Relational Schemas", Proc. of the 4th Int. Conf. on Data Engineering , IEEE Computer Society Press, Feb. 1988, pp. 276-284.

17. A. Motro, "Superviews: Virtual Integration of Multiple Databases", IEEE Trans. on Software Engineering, SE-13, 7 (July 1987), pp. 785-798.

18. L. DeMichiel, "Resolving Database Incompatibility: An Approach to Performing Relational Operations over Mismatched Domains", IEEE Trans. on Knowledge and Database Engineering, 1, 4 (Dec. 1989).

19. A.P. Sheth, J.A. Larson, A. Cornelio, and S.B. Navathe, "A Tool for Integrating Conceptual Schemas and User Views", in Proc. of the 4th Int. Conf. on Data Engineering , IEEE Computer Society Press, 1988, pp. 176-183.

20. E. Szeto and V.M. Markowitz, "ERDRAW 5.2. Reference Manual", TR LBL-PUB-3084, Lawrence Berkeley Laboratory, 1993.

21. V.M. Markowitz, J. Wang and W. Fang, "SDT 6.2. Reference Manual", TR LBL-27843, Lawrence Berkeley Laboratory, 1993.

22. V.M. Markowitz, S. Lewis, J. McCarthy, F. Olken, and M. Zorn, "Data Management for Molecular Biology Applications: A Case Study", Proc. of the 6th Int. Conf. on Scientific and Statistical Database Management, June 1992.

23. M. Zorn, "GenomeBase: Schema for a Mapping database", TR LBL-30666, Lawrence Berkeley Laboratory, May 1991.

24. O. Ritter, "The Integrated Genomic Database Project", Internal Report, Deutsches Krebsforschungszentrum, July 1992.

25. I-Min A. Chen and V.M. Markowitz, "The Object-Protocol Model", TR LBL-32738, Lawrence Berkeley Laboratory, 1993.

THE ACEDB GENOME DATABASE

Richard Durbin [1] and Jean Thierry-Mieg [2]

[1] MRC Laboratory of Molecular Biology, Hills Road,
Cambridge CB2 2QH, UK (email: rd@mrc-lmb.cam.ac.uk).

[2] CNRS Physique Mathematique and CRBM, BP 5051
34033 Montpellier, France (email: mieg@kaa.cnrs-mop.fr)

INTRODUCTION

Systematic genome mapping and sequencing projects are generating resources that will permanently change the practice of molecular biology. To maximise their effect, we have to make the information available to the scientific community in as useful a form as possible. It has been said that the sheer quantity of genomic information that we are just now beginning to gather will cause problems for any database system that must store it. That is not in itself strictly true; in fact the current total of genome mapping and sequence data, for all organisms combined, would sit comfortably in a one gigabyte disk, which is small for a workstation, and even conceivable for a PC. Furthermore, although the amount of genome data being collected is undergoing exponential growth, so is the capacity of computer storage systems, with an even shorter doubling time, so the issue of raw storage capacity is becoming progressively easier.

However, in another sense, the fear of data overload has some justification. This is because the issue is not just one of simple storage, but of integrating the new systematic data with all our accumulated experimental knowledge from classical and molecular genetics, so as to be able to select what is relevant for each scientific question. Until recently, with the results disseminated by standard scientific publication and discussion, this process of integration has taken place in the minds of the researchers. Even if all details could not be followed, the salient facts involved with some specialised function could be managed. It is this storage medium, the human brain, that is incapable of handling all the genomic data, not the computer.

Clearly what is required is a database system that, in addition to storing the results of large scale sequencing and mapping projects, allows all sorts of experimental genetic data to be maintained and linked to the maps and sequences in as flexible a way as possible. Since this is a new type of system, it seems very desirable to have a database whose structure can evolve as experience is gained. However this is in general very difficult with existing database systems, both relational, such as Sybase and Oracle, and object-oriented, such as Object Store.

We were faced with these issues three years ago, when starting a pilot project for obtaining the complete genomic sequence of the nematode C. elegans. There were dual needs: first for a system in which to maintain data for internal purposes, and second for one in which to make it public. We wanted to build on previous experience gained while building the physical clone map for C. elegans, which had been done using the program CONTIG9 [1]. An

Computational Methods In Genome Research,
Edited by S. Suhai, Plenum Press, New York, 1994

adapted form of this program, called PMAP, was made publicly available to the C. elegans research community, together with regularly updated copies of the in house data. This rapid and complete access to the map, even in incomplete form, proved to be extremely popular and successful, soon becoming a crucial resource when cloning worm genes. It therefore seemed sensible to extend the same approach, and develop a single database to hold sequence, physical and genetic map, and references, that we could use in house, and that we could distribute in read-only form freely within the worm community.

This led directly to the database program ACEDB, which is described in this chapter. Rather than being limited to the specific data that we could envisage when we started, we decided to write a general database management system that would allow easy and frequent extension and adaptation of the database schema as the project developed. For this reason, it has been comparatively easy to adapt ACEDB to be used by other genome projects working with other organisms. At the time of writing (March, 1993) there are public databases for the model plant Arabidopsis thaliana, and the mycobacteria M. leprae and M. tuberculosis, which are the pathogens for leprosy and tuberculosis. Several other databases for public distribution are under development. ACEDB is also being used internally at several sites, for example for storage of physical mapping results from human and Drosophila projects. Finally, it is being used as one of the core pieces of software in the IGD (Integrated Genomic Datatape) project (European Data Recource for Human Genom Research, Heidelberg), which plans to bring together all public human genome data in an integrated genome database. ACEDB is both being used as the primary graphics front end of IGD, and as one of the alternative back-end data storage systems.

OVERVIEW OF FEATURES

In this section we will give a brief overview of ACEDB as a biological user sees it. Overall the program is very graphical. It works using a windowing system, and presents data in different types of window according to the different types of map. The maps and other windows are linked in a hypertext fashion, so that clicking on an object will display further information about that object in the appropriate sort of window. For example, clicking on a chromosome displays its genetic map; clicking on a gene in the genetic map displays text information about the gene's phenotype, references etc; clicking on a clone displays the physical map around it; clicking on a sequence starts the sequence display facility.

The internal structures of the system, which are more general, and which contain some of the more novel features, will be described in later sections. There are interactive tools for many of these more general features available to the user as part of the graphical interface, but discussion of these will be delayed until later. In this section we will just briefly describe the windows used to display the different classes of object in the database.

MAIN CONTROL WINDOW

There is one window that is always present while running ACEDB: the main control window. This contains a list of the different available classes of objects, such as Clones, Sequences, Genes, Papers, Authors etc. To use the program in its most simple fashion the user selects a class with the mouse and types the name of the object in the yellow text entry box, then hits the return key, at which point the object will be displayed in another window. If a name template is given using wildcard characters (such as '*') then a list of all possible matching names is given, from which the user can select. There is also a menu accessible with the right mouse button that provides access to more complex features, such as the query system discussed later.

46

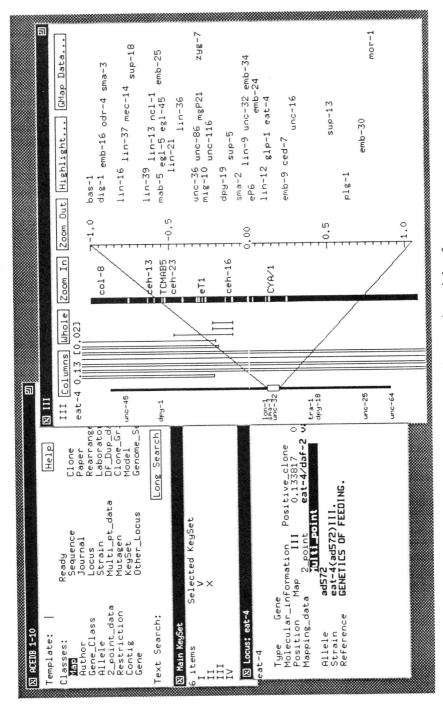

Figure 1. Main window, Genetic map, and text window of one gene

GENETIC MAP

The genetic map display gives access to genes and rearrangements on the scale of a chromosome. In the C. elegans map the units used are centimorgans. The map is displayed vertically, as are many ACEDB maps. On the left side, acting like an annotated scrollbar, is an image of the whole chromosome with a green cursor region indicating the currently displayed region. This area is zoomed up to fill the rest of the display, which consists of various zones, including from left to right: space for chromosomal rearrangements (deficiencies and duplications), an indication of physically mapped regions, a scale bar, and the genes themselves. The display can be easily scrolled up and down, and zoomed in and out using either the scrollbar cursor, or buttons that can be pressed with with the mouse. There are also buttons to allow genetic mapping data to be displayed graphically. Recently we have extended the mapping data package to also allow calculations on mapping positions to be made from the map data.

If the user double clicks on any item a new window pops up with text information about the object. This text information is layed out hierarchically, in what is called a tree structure. The section below on "Organisation of data" describes further this tree structure, which in fact is the primary way of storing information in ACEDB. The maps are merely derived from the data stored in tree form with each object.

PHYSICAL MAP

This is the primary display for looking at the positions of clones within clone contigs. It is (currently) a horizontal map, based very much on the map display of CONTIG9 [1]. Again at the bottom is an annotated scrollbar showing a wider region of the contig. At the top are a number of sections showing different types of clone, i.e. probes, YACS, cosmids. Under these are spaces for genes that have been attached to the map by localising them onto a clone, and remarks, which can be freely attached to any clone. Also in this region is a green horizontal band linking back to the genetic map. Again, whenever an item is selected, all the related items are also lit up; e.g. selecting a gene might show the clone that contains it, and any remarks attached to the clone. If the item is double clicked, then text information about the item is displayed.

HYBRIDISATION GRID

This window provides access to one of the forms of raw data used in building the physical map, which is hybridisation of probes to a grid of clones arrayed on nylon filter. The grid is displayed schematically, and the hybridisation pattern can be entered by clicking on the appropriate squares. Once this is done, the inferred loci on the physical map are determined automatically. There are also facilities for comparing one pattern with another, and for displaying the results of both real and hypothetical pooled probings. This tool was used to position 1100 nematode cDNA's on the C. elegans physical map, and is also used by a human mapping project for data acquisition.

DNA SEQUENCE

The sequence display function in ACEDB is particularly flexible. As well as being able to display the actual nucleotide sequence and protein translations of it in all frames, there are a wide range of different schematic display options available. These allow features to be shown

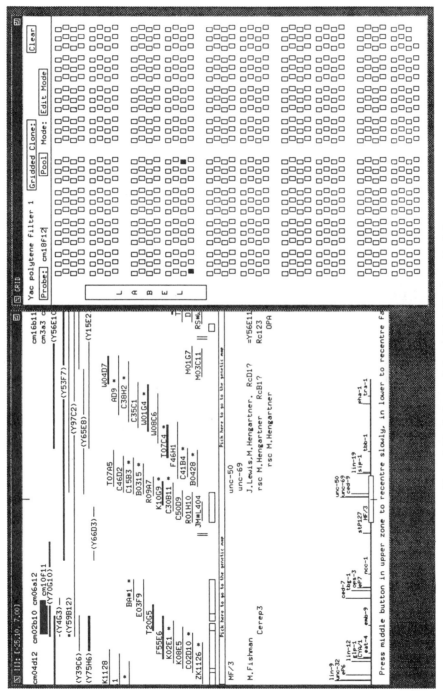

Figure 2: Physical map window and hybridisation grid window.

49

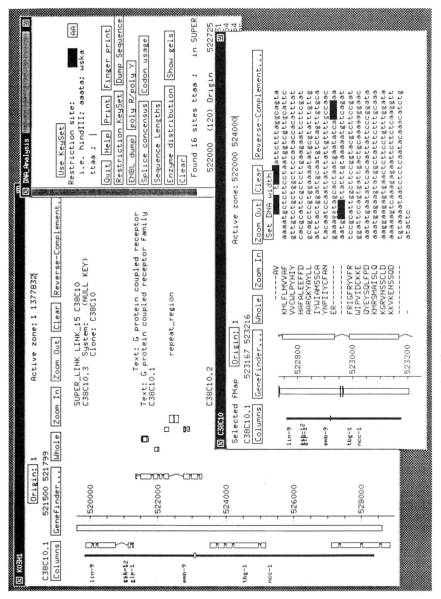

Figure 3: Sequence windows.

with at many different scales, with simplified results on a megabase scale at one extreme, and details of short tandem repeats at the other. Selection of exactly which options are shown is controlled by a separate menu.

As well as displaying annotations and precalculated information, the sequence window supports several types of calculation. In particular there are a range of facilities derived from the Genefinder program (Green and Hillier, personal communication) for predicting gene structures in genomic DNA sequence based on likelihood predictions of splice sites and codon usage. These are in fact used to annotate the nematode genomic DNA sequence before submission to EMBL. There are also restriction site detection tools, and tools for extracting subsequences and translations of predicted genes.

ORGANISATION OF DATA

Definition of the schema

ACEDB is an object-oriented system. This type of organisation of data is much more intuitive than one based on relational tables. Because of this it permits more direct input from working biologists on the construction and refinement of the schema, or data structure.

Each object is represented by a unique identifier, its name, which is followed by an ensemble of attributes organised into a tree. The nodes at branchpoints of the tree are all named. The branches typically terminate in pointers to other objects, or data, which are numerical values, character strings. A bare branch ending just in the named branchpoint can be used to indicate presence of a binary property. There is also the possibility of constructed subobjects, similar to expanding a leaf node in place recursively into a full object with its own branches, rather than maintaining merely a pointer to an external object. Arbitrary text comments can be attached freely at any point in the tree.

The objects are allocated to classes. Each class has a model, specifying the maximal extension of the branches, and the types of data or classes of pointer permitted at each position. Individual objects, which are instances of the class, in general only have a part of the branching pattern permitted by the model. This approach gives a triple advantage:

a) Poorly studied objects, which are by far the most numerous, take little space on disk and in memory, which strongly increases the speed and efficiency of the program.

b) If one wants to extend the schema, which we do frequently, all that needs to be done is to add another branch to the model. Of course none of the existing objects contain this branch, but they remain valid because there is no requirement for objects to contain any particular branch of the model.

c) The ability to add personal comments that are ignored by searching algorithms allows flexible annotation of data sources, reservations etc. without affecting internal search procedures.

Here is a part of the definition of the class Gene, and an example of a gene:

```
?Gene     Reference_Allele ?Allele
          Molecular_information    Clone    ?Clone XREF Gene
                                   Sequence ?Sequence XREF Gene
          Map     Physical pMap UNIQUE ?Contig XREF Gene UNIQUE Int
                           Autopos
                  Genetic  gMap ?Chromosome XREF Gene UNIQUE Float UNIQUE Float
                  Mapping_data    2Point ?2_point_data
                                  3Point ?3_point_data
          Location ?Laboratory #Lab_Location
?Lab_Location    Freezer        Text
                 LiquidN2       Text
```

```
ced-4    Reference_Allele n1162
         Molecular_information   Clone   MT#JAL1
                                 Sequence ?Sequence XREF Gene
         Map     Genetic  gMap III -2.7
                 Mapping_data    2Point "ced-4 unc-32/+ +"
         Location Cambridge Freezer A6
```

Note that each object belongs to just one class. We have deliberately chosen to avoid multiple inheritance. This concept is at the same time notoriously difficult to implement efficiently and very difficult to use, because the inheritance graph easily becomes encumbered with potential conflicts amongst super-classes.

Our alternative to multiple inheritance is to restrict the number of classes, but allow a wider variety of objects within a class. In this system, it is possible that two objects in the same class have few or no branches in common. For example consider two genes, the first studied by classical genetics and uncloned, the second cloned by similarity to a protein in another organism. These objects could be considered as archetypes of two subclasses of the class gene. But such simple cases are relatively rare, a third gene could have data for some fields of one type, and some of the other, and one is rapidly led to a combinatorial explosion in the number of classes. Our approach lets us capture without difficulty all the intermediate cases, and we only need around fifty classes to hold nearly a hundred thousand heterogeneous objects.

As well as the classes of tree objects described above, also denoted type B classes, we have type A classes, which contain general arrays of data, and which allow more rigid but more efficient storage of data such as DNA sequences.

The schema itself is stored in objects within the database, allowing a simplified startup procedure and dynamic editing of some features.

Ace files: an ascii edit language

Although data are stored internally in a binary form of the trees discussed above, they are normally entered via simple ascii files known as ace files.

Each paragraph in these files corresponds to one object, and must be separated from the next by one or more blank lines. The first line indicates the class and name of the object to be created or edited. Following lines start with the name of a branch node, followed by numerical or text data, or names of other objects to be pointed to. They are interpreted according to the model. Keywords such as -D or -R specify actions to be taken, with the default action being to add the data into the database. As in C++, // indicates a comment in the file.

Example:

```
// First let us define a sequence:
Sequence ACT3
Title ``C. elegans actin gene (3)''
Library EMBL CEACT3 X16798

// next the corresponding DNA (A class with special reader)
DNA ACT3
aagagagacatcctcccgctcccttcccacacccacttgctcttttctat
tgaccacacattatgaagataaccatgttactaatcaaattcgtgttctt
ttccaatttctttttc

// here we change the name zk643 (if it exists)
-R Sequence zk643 ZK643         // R for ``rename''
// here we change one of the authors of the paper [wbg101]
Paper Nurture:7:234-242
-D Author ``Kimble JE''          // deletion of Kimble
Author ``Ahringer JA''           // addition of Ahringer
```

It is specifically because the objects have a public unique ascii identifier, the doublet [class:name], that these edit commands are well-defined and can refer to precise objects. If the object is not known yet, it is created, else it is modified. If a delete or rename operation finds nothing to delete or rename it moves on silently. If an instruction makes no sense according to the model, for example by referring to an unknown branch point, the user is warned and the paragraph is skipped. Together these properties also allow repeated reading of the same file without changing the database contents, and transfer of information between databases that may not match exactly, something which is very hard with traditional database systems. Indeed they even allow transfer of commonly meaningful data between systems whose schemas differ.

As well as reading in data, we can also export a set of objects in ACE file form. An external program, acediff, takes as input two such ace files, and generates a third that would have the effect of transforming a database containing data as in the first file into one containing data as in the second file. This program can be used to generate update files for remote copies of a central database, and in fact this is the procedure we use to distribute the nematode genomic database. There are also facilities within ACEDB for certain types of specialised data output, such as DNA in FASTA format.

THE DATABASE KERNEL

We wanted ACEDB to be portable and efficient. The system is therefore built using the standards of C, Unix, X11 and PostScript. Because we ended up defining a new type of database system we had to write our own database management system. We also wrote our own graphical library on top of X, which can be reimplemented using other underlying graphical systems (e.g. Apple Toolbox), and a number of macro-based extensions to C that seemed useful to us. ACEDB is therefore not tied to any particular machine, nor operating system, nor even (after a little work) windowing system. It contains an internal help system, and a crash recovery system (when possible). Some of these functions are described in this technical section.

Disk storage

The conceptual unit of transfer between disk and memory is the object. Since the objects are trees or arrays of arbitrary size, we wrote a relatively complex module to pack them into and out of fixed size blocks (one or two kilobytes). The many small objects are brought together in groups that each fill one block; large ones are split over several blocks. The system is speeded up by two levels of internal cache, the first acting on the fixed size blocks, the second on developed trees. These both work as last referenced/last out queues.

Since the class of an object is known at all levels, it is easy to selectively optimise the storage of certain classes, as we do for example with DNA.

Within any particular session, all modified objects are rewritten to new disk locations, which allows us to store multiple versions of an object, and recover from crashes by going back to the last verified save state of the database.

Indexing system

Each object is identified externally by the ASCII doublet [class:name]. Internally it is represented by a unique 32 bit key, 8 bits being used for the class, and 24 for the location within the class. Linking these are hash tables that map the known names of each class to keys. For each key there are then a set of indices containing the disk address of the object, cache addresses if it is in memory, flags indicating its edit state and other properties, a pointer back to its name, etc. There are separate index and hash tables for each class. These are loaded into memory as needed, and take up about 30 bytes per object.

Only one user at a time is permitted write access. The set of changes made until write access is given up constitutes a session. When a session is saved, first the changed objects are flushed to disk, then the indices and hash tables for any altered classes are written (as type A class objects, and hence also to new disk locations). Finally a pointer in the superblock is changed to point to the new index information. Once this is done the system will start up with the new indices. Any crash before this point will leave the system so it starts up by retrieving the old indices, and hence the old objects from before the aborted session.

Data access library

All simple access to data within the program uses a subroutine library that allows access via the names of terminal branchpoints, rather like with the ace files. There are two steps: the first recalls an object from disk given a key, and returns a handle; the second uses this handle to recover data. Since any data may be missing, these routines all return a boolean value indicating success or failure. As an example, to recover the date of a paper, one might write (in C code):

```
void paperDate (KEY paper)
{ int year; OBJ obj;
  if ((obj = bsCreate(paper)) && bsGetData(obj,_Year,&year))
    printf (``Paper %s was published in year %d\n'', name(key), year) ;
  bsDestroy(obj) ;}
```

Query language and Keysets

As well as accessing data via the direct subroutine package, there is a powerful and general query language that allows higher level manipulation of sets of objects, which are known as keysets. The basic operations in the language are (1) perform filtering operations on a set of objects, either on the basis of their names or the data they contain, and (2) to follow pointers to retrieve other objects. These operations can be combined using boolean operations into complex query sentences.

The resulting keysets can be used in in various ways: single items can be looked at interactively, the whole set can be a starting point for further queries, it can be dumped out as ascii ace file (see above), or it can be saved in the database with a user-specified name. Boolean set operations can also be used to combine sets. An important feature is that sets can contain objects from many classes. One example of how this is used is via another query operation, "text search". This performs a search on all text stored in the database, and returns a list of all objects that either have names matching the search string, or contain text that matches it. For example a search for "muscle" might return genes with muscle phenotypes, papers with "muscle" in the title, sequences of muscle proteins, etc.

The query package is available both to the user, via an interactive interface that allows saving, recovery and reuse of queries, and to the programmer via a library of subroutines. In fact the main control window is implemented by setting up a limited set of straightforward queries.

There is another facility for general data presentation based on the query package, called the "Table Maker". This allows the user to construct tables of displayed information in a similar way to using a spreadsheet. The difference is that, in the table maker, new columns are derived from previous columns by queries, not by calculations. Once again, the instructions for defining a table can be built up interactively, and stored in a file once they are correct.

Textace and the server/client architecture

All the discussion so far has concerned the graphical version of ACEDB that most users see. However there is also a text version of the program, textace, that can be run from the Unix command line. This basically contains the kernel with a simple command parsing interface.

Although it can be useful for extracting data over a serial line, the main purpose of this version is to enable a client/server architecture for ACEDB.

The way this is done is for the server to contain a full copy of the database, and for the client to start with an empty database. When a query is generated by the client, the server resolves it and sends the result back to the client as an ace file, which is parsed into the client database in the normal fashion. In fact the server is acting in exactly the same fashion as an independent textace program, except that it is connected to the client by a pair of sockets, rather than by the standard I/O. This type of structure is only possible because ACEDB allows meaningful data transfer between non-identical databases, via ace files. The client database can either be allowed to accumulate during the session, acting as a local cache, or can be restricted so that all calls for data are resolved by passing them back to the server. Of course the former can become much more efficient, while susceptible to data becoming stale if it is edited by another process. It is clear that when editing data, the objects must be retrieved from the server and locked there, rather than updated based on a local copy.

CONCLUSION

ACEDB is publicly available by anonymous ftp, both in binary form and as source files. There are three primary ftp sites: cele.mrc-lmb.cam.ac.uk (131.111.84.1) in England, directory pub/acedb; lirmm.lirmm.fr (193.49.104.10) in France, directory genome/acedb ncbi.nlm.nih.gov (130.14.20.1) in the USA, directory repository/acedb. In each case the file NOTES gives further instructions on retrieving the program. The data for the current version of the C. elegans database is available in the same directories.

Although we make the source code available (under a licence restricting commercial exploitation), and we encourage development of new specific application code, we hope that the community of groups using ACEDB can keep to a single database kernel. This can be achieved by establishing good contact between groups that are doing development work, and folding kernel changes back into the official release version described above. With this policy, we believe that ACEDB can continue to support a growing community of genome database providers, covering many different genome projects.

REFRENCES

1. "Software for genome mapping by fingerprinting techniques", J. Sulston, F. Mallett, R. Staden, R. Durbin, T. Horsnell and A. Coulson, Comput. Applic. Biosci. 4, 125-132 (1988).

THE INTEGRATED GENOMIC DATABASE (IGD)

Otto Ritter

European Data Resource for Human Genome Research
Department of Molecular Biophysics
Deutsches Krebsforschungszentrum (DKFZ)
Heidelberg, Germany

1. ABSTRACT

This paper presents an overview of the design and implementation of the prototype version of the Integrated Genomic Database (IGD) [1]. IGD should provide for a general distributed system to access information of interest for genome researchers, human geneticists, and molecular biologists. It aims to integrate the genetic data in a comprehensive database, and provide end users with a set of tools to retrieve, display, edit and analyse the data in one framework. The prototype system integrates public data of several major databases physically into a server read-only database. IGD front-end clients then allow scientists to store retrieved data in a local database, where these can be commented or edited, and, if needed, merged with local experimental data.

IGD does not attempt to replace any existing or potential future data resources. Instead, it manages communication between these resources, and provides software tools to interface them in a uniform fashion.

2. INTRODUCTION

Informatics has become an essential tool in genome research. Growing volumes of raw and derived data are organised into numerous database systems and analysed by a variety of software tools. To retrieve and process any online available information, scientists have to invest extensive efforts to learn the idiosyncratic use of all different database interfaces, query languages, and parameter specifications for individual analytical programs. Even more efforts have to be invested into building laboratory information management systems. Further on, databases and analysis tools are mostly incompatible in their data formats [2].

To optimize the utility of accumulated information and analysis resources, all the databases and tools should be interoperable, in the ideal case. To date, there is no satisfactory integration even on the data level. If we want to understand the organisation and functioning of genomes of higher organisms, we need to locate the experimental and conceptual entities (genes, DNA markers, genomic clones, regulatory elements, binding sites, etc.) to defined parts and regions in the genomes. Maps of different sorts and scales are therefore constructed and, ultimately, combined together. The location information is equally important as the sequence and structure information. To get the best for our analyses, we need an integration of the sequence information, molecular structure, genetic and physical location, variation information, and the phenotype and population information. It is the primary objective of the IGD Project to build such an integrated information management environment.

There is no database available that would provide for a comprehensive integrated view of the human genome. There are public consensus databases like the Genome Data Base (GDB) and the Online Mendelian Inheritance in Man database (OMIM), experimental databases like the Reference Library Database of ICRF (RLDB), the UK Probe Bank, the cDNA database, the CEPH/GENETHON database, and many others. Then there exist organism-nonspecific databases like the Protein Data Bank (PDB), the GenBank and EMBL nucleotide sequence databases, the Swiss-Prot and PIR protein sequence databases, and many subject databases derived (and extended) from mainly the general sequence databases: e.g., the EPD, REBASE, TFD, PROSITE, etc. Recently, integrated genomic databases are being developed for some model organisms, eg. GBASE and The Mouse Encyclopedia for the mouse, ACeDB for the nematode C. elegans, FlyBase for Drosophila, PIGMAP and BOVMAP for pig and cattle, PGD for several plants, AAtDB for Arabidopsis, etc. A good review of the current molecular biology databases can be found in [3].

Some integration efforts for human (or general) genome data sets are under way. GenInfo and Entrez at the US National Center for Biotechnology Information (NCBI) [4], The Chromosome Information System at The Lawrence Berkeley Laboratory [5], the Reference Library of ICRF [6], the Contig Browser at the Lawrence Livermore National Laboratory [7], the experimental databases developed at CEPH/Genethon in Paris, and the IGD at the German Cancer Research Center (DKFZ), Heidelberg, offer so far partial integration. Several projects concentrate on providing a user friendly interface to genomic databases: Genome Topographer at The Cold Spring Harbor Laboratory, GnomeView at The Pacific Northwest Laboratories, Richland, GDB-lite at Baylor, GenoGraphics at The Argonne National Laboratory. And, of course, most of the above mentioned integration efforts work on their graphical interfaces, too.

What software is actually used for data management? The genomic database projects use either well established commercial database management systems (relational like Sybase, Oracle, and Ingres, and object-oriented like GemStone and Object Store) or they use application-specific software for their data management. The ACEDB software [8], originally developed for the nematode genome project, has become a system of choice for several other organisms [9], and is being used as a general genomic database management system. Some genome related data sets are maintained as mere collections of text files, too. With some simplification we can state that today the major genome databases are either relational or ACEDB-based.

There are, in principle, two possible approaches towards database integration: virtual and physical. In the virtual case, a smart system only pretends to be a physical database - it takes the question (query) and splits it into subqueries, asks then all the appropriate individual databases to process their subquery, and, finally, assembles the response on the fly. The obvious advantage of a virtually integrated system is that it is always up-to-date and doesn't occupy much space. On the other hand, such a system has to wait with the answer until all the subqueries are successfully processed, and this may become a bottleneck when one of the target databases is not reachable due to either network or host-computer problems.

A physically integrated database periodically imports all data from the individual databases, collates them, and stores them in its own DBMS. The data importation periods may vary for each target database, depending how often these are updated, and how much we allow the integrated database to be "outdated". It is not necessary to split the query, and it is not necessary to establish connections with the individual target databases at query time. For these reasons it may seem that physically integrated systems perform better than the virtual ones. Well, not always - don't forget that the virtual system processes the query in a parallel fashion on several independent processors!

For the IGD prototype physical integration was the easier and more robust choice: disk space is relatively cheap, there is no need for absolute data synchronisation, and the wide area networking situation in Europe is not yet good enough to allow for a reliable and fast operation of a very large virtual database. Nevertheless, IGD anticipates a future switch from the physical to the virtual mode.

The following three observations have played decisive role in the choice of a DBMS for IGD:

- Relational systems (RDBMS) are based on a standard data model and have a standard language, the Structured Query Language - SQL [10], for data operations. They are robust and reliable, provide for query optimization and transaction management, and are sure to be supported and further developed. On the other hand, they are more suited for business rather than scientific applications.
- The ACEDB system has an object-oriented data model and a very appealing and intuitive user interface with graphical display methods for genetic and physical maps, sequences, clone grids, etc. It is available at no cost and is highly portable over UNIX platforms with X Windows, but its query language is less powerful than SQL, and it doesn't support hierarchical data access and transaction management. Also, ACEDB doesn't yet have the full capability to specify data integrity constraints.
- Commercial object-oriented or extended-relational DBMS might be well suited for the management of complex data, but they do not provide graphical interface tools comparable to those of ACEDB.

We have decided to complement the powers of the ACEDB and Sybase database management systems, and to use them both in parallel for the physically integrated genomic database server. We've further decided to modify the role of the IGD database clients - instead of being only data browsers, they get their own DBMS to store the retrieved data. Obviously, the choice here was the ACEDB. More on this follows in the design and implementation sections.

The rest of the article is organised as follows. Section 3 gives an overview of the whole system design, and the functional specifications of individual system components. Also the overall objectives are stated here. Section 4 describes the prototype implementation of the designed system, and Section 5 discusses the prototype results and future plans.

3. SYSTEM DESIGN

3.1 Divided roles

It wouldn't make much sense to put all the desired functionality into one single program - such a program would have enormous requirements on computer memory and performance, and it would be very difficult to maintain. The trend in software development today is to split complex functionality into modules which communicate and work together. These modules pass and receive messages (requests, data, signals, etc.) in agreed format, but they don't "see" how these messages are processed internally by the other modules. Ideally, it shouldn't matter whether this communication takes place on one computer or over a network.

Following these concepts, IGD is designed as a network system based on a client/server architecture. This means that the application (e.g. user interface) is separated from the back-end database server. Application programs, called clients or front-ends, run on computers local to the user, and they communicate with other programs, called servers, running mostly on remote computers connected by a network. In the traditional client/server model some processes, the clients, are only requesting resources while other processes, the servers, are only providing

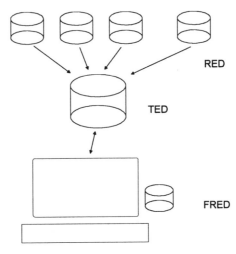

RED

TED

FRED

Figure 1. Basic layers of the IGD system.

those resources. In IGD particular processes act both as clients and as servers (e.g. when two front-ends exchange data). The whole system is open: new clients and servers can be added as far as they comply with the agreed communication protocol and query/data format.

With regard to the origin and scope of data, the system can be subdivided into three functional levels:

1) resource databases which contribute data
2) target database servers which manage the integrated data
3) front-end clients which manage data locally to the user

In addition, there is the fourth logical level of communication tools which transfer and transform the queries and data. The four levels are implemented using four relatively independent subsystems: **RED** (the Resource-End Database), **TED** (the Target-End Database), **MIM** (the MIddle Manager), and **FRED** (the FRont-EnD interface) (see Fig. 1).

REDs are processes that take updates from the resource databases (one RED per each such a database), transform the data into suitable format, and send them over MIM to the TED where they are read into the database.

TED runs a database server with an interface to the MIM. Any UNIX filter that takes a query in a language and format supported by IGD on its standard input, puts suitably formatted data onto its standard output and the error and diagnostic messages onto its standard error ouput, may serve as an IGD TED. Of course, it only makes sense if its data model is equivalent to the integrated model or is its submodel. TED runs in a read-only mode with respect to the FREDs, and in a write-only mode with respect to the REDs.

FRED is designed as a set of programs that users are running on their computers. (see Fig 2.) These programs communicate with each other and also with the remote parts of IGD. The communication with remote resources is primarily asynchronous so that the user is not blocked waiting for answers to his/her queries. FRED assists the user in composing queries in a high level object query language which gets translated by IGD into the language of the target database, retrieves the data from the outside world, stores them in a local database, displays and manipulates the local data, and enables to call services which might be distributed over the network. Users see FRED as a hierarchy of manager programs. There is the Query Manager to take care of queries (queries can be saved in a database, searched for, retrieved, and reused), and there is the Response Manager to retrieve and process incoming messages, mostly replies to earlier queries. Data retrieved from remote servers (TEDs) or elsewhere (files) may be uploaded into the local database. The Communication Synopsis manager gives an overview of

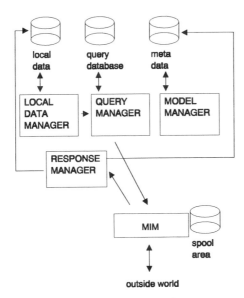

Figure 2. The IGD FRED environment.

the incoming and outcoming messages and their status. The Local Data manager calls basically the local database, i.e. ACEDB at the moment. The Model Manager displays the IGD model as an EER diagram and browses through data classes and their attributes. Authorized users may edit the model and backtranslate it into an ACEDB model specification. This mechanism enables to modify the FRED's view of IGD data and to merge the public model with local models.

3.2 Conceptual database schema

Whichever approach to the integration we take, virtual or physical, we always have to provide a conceptual schema of the integrated database in terms of a data model, and we have to provide some query language on top of that schema. The most challenging and not always satisfactorily solvable problem in the schema design is how to map the semantics of the individual database schemas onto the integrated schema.

The main data objects in IGD are records of experimental entities (clones, probes, libraries, etc.), records of experimental processing (library construction, probe hybridization, sequencing, electrophoresis, etc.), records and results of analytical processing (protocols, mapping methods, genetic maps, physical maps, sequence alignments, etc.), and records of relationships between them. Other data objects are sources (people, laboratories, bibliography) and cross-references to objects in external databases (sequence, phenotype, location, annotation).

Major principles driving the schema design of IGD were:

- flexibility of data representation
- extensibility of data representation
- data protection (hierarchical access)
- allowance for conflicts and variation
- allowance for complex relationships
- distinction between experimental facts and analysis results
- traceable history of the database objects
- access to the data only through a high level language
- provision of an application program interface to the database
- provision of interfaces with local database management
- use of standards in data representation and exchange

Translations	FRED	Intermediate level	TED

Model / Schema description language	object-orientied: data trees with logical pointers and up to two-dimensonal attributes plus display methods.
	
	`?Gene    Name`
	`     Gene_class UNIQUE`
	`       ?Gene_Class XREF`
	`                   Gene`
	`   Clone ?Clone XREF`
	`                 Gene`
	
	`?Clone   Gene    ?Gene`
	`         XREF Clone`
	`         Locus   ?Locus`
	`            XREF Clone`
	`         Remark  Text`
	
	Size: 25 KByte (models.wrm)
		Extended Entity Relationship Model (EER): with inheritance, ID-dependency and emulation of multi-valued attributes 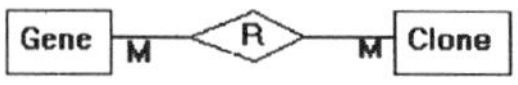
		
		`Gene` M ◁ R ▷ M `Clone`
		
		Size: 80 KByte (.sdt file)
			* relational: set of tables in 3NF plus triggers enforcing referential integrity
			* flat file: set of text records consisting of fields
			
			`create trigger`
			`insertGene_Clone_P`
			`on Gene_Clone_P`
			`for insert as`
			`begin`
			`...`
			`        select @ins1Gene`
			`= count(*) from`
			`inserted, Gene where`
			`inserted.Gene_objID =`
			`Gene.objID`
			`        if @null1Gene +`
			`@ins1Gene +`
			`@null2Clone + @ins2Clone`
			`!= 2 * @row`
			`raiserror 70001 "Cannot`
			`insert ..." end`
			
			Size: 1850 KByte (!)

Query language	- Graphical and Text through ACeDB
	- COQL
	- IRx (efficient text retrieval)
	- SQL (direct access to TED)
	- free format (simple text search)
		COQL (Concise Object Query Language) for EER models: example:
		
		`CONDITIONS`
		`   Chromosome:`
		`Chromosome_id [match`
		`"1"];`
		`OUTPUT`
		`   Chromosome:`
		`Chromosome_id;`
		`   Chrom_band : ALL;`
		`CONNECTIONS`
		`   Chromosome Locus gMap`
		`Chrom_band ;`
		`END`
			Query Languages of the DBMS used:
			
			SQL,
			ACeDB,
			IRx,
			free format

Data representation	.ace file:
	
	`Image : "G6PD-contig"`
	`Author "Kioschis"`
	`Author "Poustka"`
	`Gene    "G6PD"`
	`Pick_me_to_call`
	`"xv" "g6pd.comp.tif"`
	
	`Image : "gel1"`
	`Pick_me_to_call`
	`"xv" "gel1.tif"`
		virtual EER data format:
		
		`<entity-name1>`
		`   \t <attr-name1> \n`
		`   \t <attr-name2> \n`
		`...`
		`<entity-namen>`
		`   \t <attr-namex> \n`
		`   \t <attr-namem> \n`
		`<attr-val1>`
		`   \t .. <attr-valm> \n`
		`...`
			Sybase output:
			
			`last subquery ...`
			` Gene_gMap_P_gMap_7`
			`Gene_gMap_P_gMap_8`
			`------------------ --`
			`------------------`
			`            57.000000`
			`57.000000`
			`            426.000000`
			`62.000000`
			`...`
			`(1932 rows affected)`

Data base management system	ACeDB	(UNIX files and pipes)	Sybase, IRx

Figure 3. Various aspects of the mapping between Sybase and ACEDB.

4. IMPLEMENTATION OF THE PROTOTYPE

We have decided to use the commercial relational DBMS Sybase [11] at the back-end (TED) and the non-standard non-proprietory DBMS ACEDB at the front-end (FRED) of the system. One could say that from the data modeling point of view ACEDB is structurally object-oriented. It has classes and class-dependant methods, but does not support inheritance. In order to map between the Sybase and ACEDB data models, we have introduced an intermediary Extended Entity-Relationship (EER) schema level. We developed schema conversion tools to automatically translate between ACEDB and EER schemas, and we employed the ERDRAW [12] and SDT [13] tools of V. Markowitz, to translate from EER into

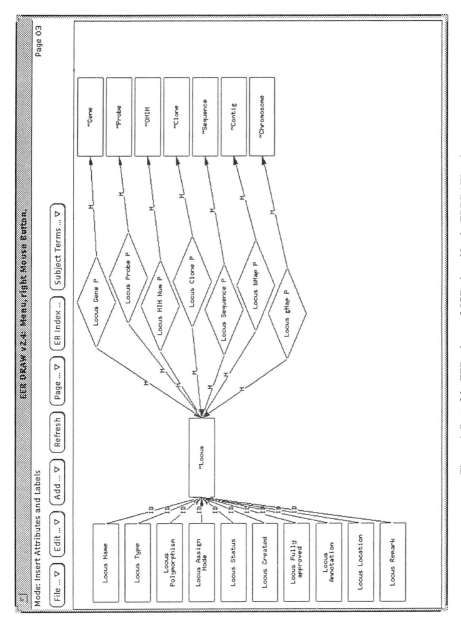

Figure 4. Part of the EER schema of IGD viewed by the ERDRAW tool.

Sybase schema specifications. Figure 3 shows very simple examples for the mapping between the FRED (ACEDB) and the TED (Sybase) database systems on three levels: translation of the schema, translation of queries, and translation of data.

For the semantic integration of the resource databases (REDs), we used the ACEDB modeling level, and we then subsequently translated the ACEDB schema into the EER and Sybase specifications. Developing the database schema from the ACEDB end is much easier and less error prone than developing the schema directly in Sybase SQL. For IGD, the ACEDB specification is captured almost entirely in one file 25 kb large, while all the corresponding SQL definitions take 1850 kb in a number of files. The interim EER representation of the schema on the way from ACEDB to Sybase is not a mere byproduct. It is itself very useful as a standard view of the schema.

We use the ERDRAW tool to present the EER schema graphically (see Figure 4) to help the user build a query in a high-level object query language, COQL [14]. COQL queries are automatically translated into Sybase SQL using the query translators of V. Markowitz.

The flow diagram of Figure 5 shows the most complex case how a query is sent from FRED to TED. It has to be said that the default loop would be much simpler - in most cases an ACEDB query to an ACEDB TED would serve the purpose without any need of query or data translations. However, due to the current limitations of the ACEDB query language, some expert users may decide to use the COQL or SQL query languages for more sophisticated analysis.

The schema of IGD was produced by matching the semantically equivalent objects and attributes of each of the REDs onto equivalent or newly introduced objects and attributes of the ACeDB schema. Semantic conflicts at the schema level were resolved by juxtaposting the different versions. This same method is used for integrating data: for each homonym all definitions are stored equally in IGD, however in a form allowing to trace the origin in each case. Figures 6 to 9 give examples how the integrated data are presented graphically.

In the main window on Figure 6 in the upper left corner genes with names starting with "pr" have been selected; the selection appers in the window below. Double clicking on the PRF1 symbol opens the right window with a cytogenetic map. PRF1 maps to the 17q11 region. Selecting the symbol again opens the lower left PRF1 window with locus data. The in-situ hybridization image gives a contradictory evidence that the PRF1 gene is located on the chromosome 10.

We have implemented the MIM communication level using the UNIX remote shell call (rsh) mechanism for the synchronous part of the communication, and file buffering for the asynchronous layer. We investigate into the more portable communication method of remote procedure calls, RPC. In the future we plan to support also non-interactive FTP and electronic mail communication layer of the MIM.

The FRED graphical user interface uses the X Windows Version 11 Release 5 windowing system of MIT with Athena widgets library (see Fig. 10 for an example of the FRED look and feel). FRED processes use UNIX semaphores and signals for inter-process communication. All FRED processes, including the ACEDB database manager, are written in the C language. The various TED processes are written as simple UNIX filters; they are shells encapsulating possible query transformations, calls to appropriate DBMSs (e.g. Sybase isql, IRx server, ACEDB tace server), and calls to data convertors. To export data from REDs, convert them and import them into TEDs, simple scripts in the GNU AWK or NAWK languages are pipelined together.

To link FRED data with processing methods we use the "PickMeToCall" mechanism of ACEDB to call external processes and pass them arguments.

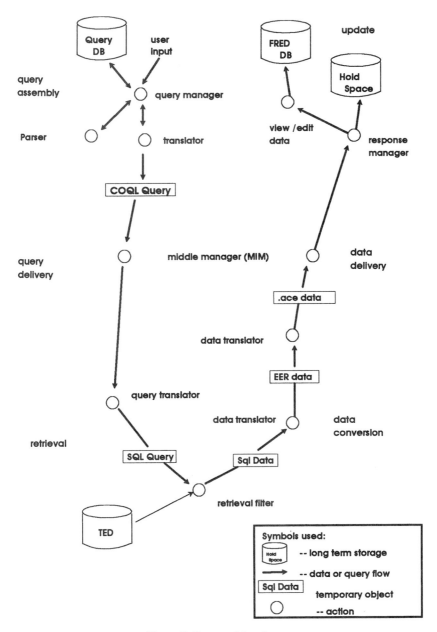

Figure 5. Query and data flow.

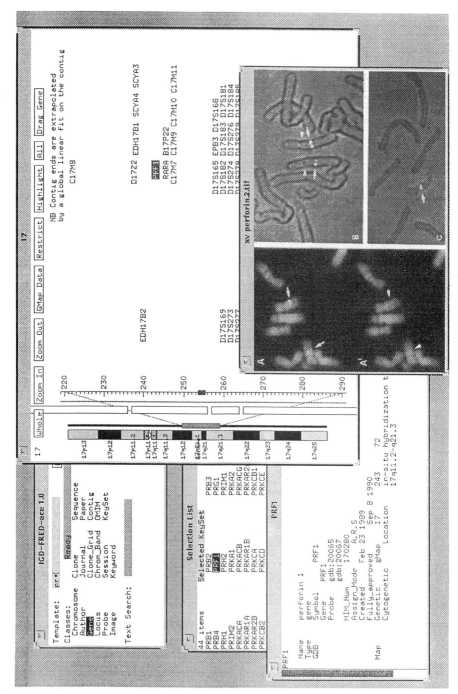

Figure 6. The Perforin 1 gene.

66

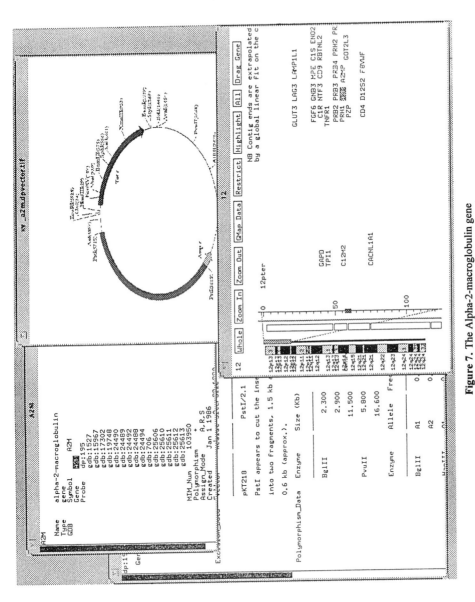

Figure 7. The Alpha-2-macroglobulin gene

The A2M window shows locus data of the human alpha-2-macroglobulin gene. Notice the **dp:1** probe from the DNA Probe Bank database; its data are given in the window below and its cloning vector map is displayed, too.

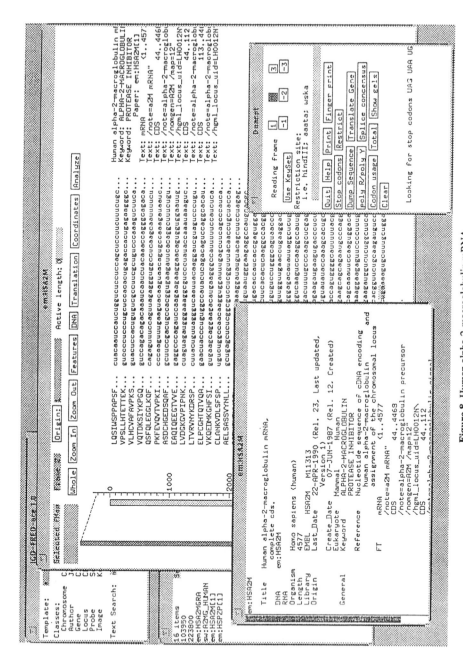

Figure 8. Human alpha-2-macroglobulin mRNA

The annotation and sequence of the EMBL HSA2M entry is shown here. Several analysis tools are callable from the lower right window.

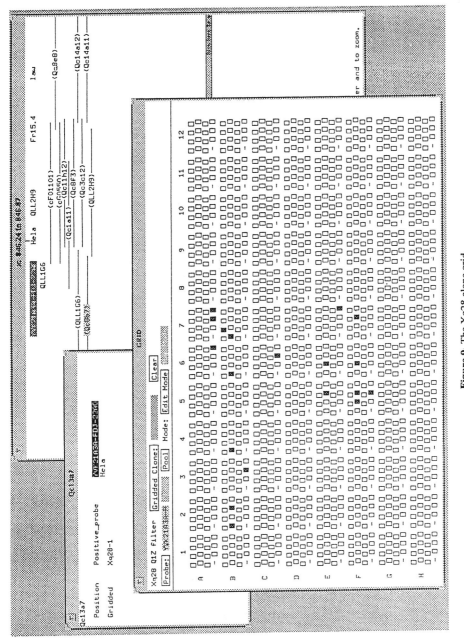

Figure 9. The Xq28 clone grid

A cosmid contig from the human Xq28 region is shown together with the hybridization of one probe on the clone grid.

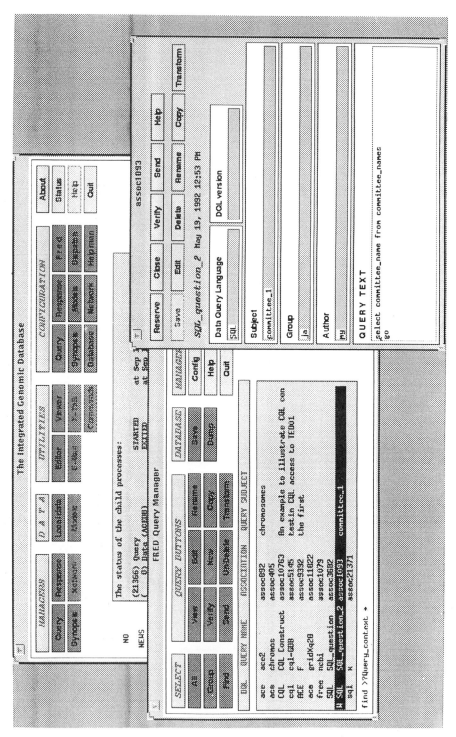

Figure 10. Query manager of the FRED

The following resource database (REDs) have been so far imported into IGD:

- The **EMBL** Nucleotide Sequence Database [15]. We have imported all primate entries into IGD, but the parsing and translating methods apply generally to the whole EMBL database. We import all the feature data on sequences, but we do not parse the feature table in detail yet. ACEDB at the FRED side supports limited DNA sequence analysis. For more complete coverage of analysis methods we plan to interface the GCG package and let the analysis be done outside ACEDB on exported data sets.
- The **SWISS-PROT** Protein Sequence Database [16]. We have imported all human amino acid sequences into IGD, but like with EMBL, we can easily import the whole SWISS-PROT using the same tools. There are no analysis methods for protein sequences available at the FRED now.
- The **Genome Data Base** (GDB) [17]. So far we've imported data on all the genes and DNA loci, probes, and literature references. There is much more in GDB, but we did not attempt to parse the whole of GDB at the first run - it would have gone far beyond the prototyping phase.
- The Online Mendelian Inheritance in Man (**OMIM**) Database [18] is fully imported. OMIM documents are linked with GDB loci in both directions. Certain fields (e.g. keywords and the clinical synopsis) are cross-referenced with the whole IGD. Full text of the documents is also searchable for user-defined patterns.
- The **Reference Library Database** (RLDB) [6]. We have imported the information on all publicly available RLDB probes and their suppliers.
- The UK **DNA Probe Bank Database** [19]. We have imported all the probe data and also graphical maps of cloning vectors.

Beside these public data sources we have put into IGD samples of fluorescent in-situ hybridization data and images, and also a dense physical map of the human Xq28 region together with detailed experimental data on the gridded clone library [20], hybridization results, etc., to demonstrate how public and private data can be merged together in one system.

5. DISCUSSION

The purpose of the IGD prototype system was to test the "RED-TED-FRED" concept, and to assess the complexity of the integration of heterogeneous schemas. It seems that the conceptual design has been successful. The implementation has shown that, although functional and flexible, the user interface is too complex and intricate to be useful to biologists without considerable computer experience.

Now, together with Martin Bishop, Steve Bryant, Richard Durbin, Hans Lehrach, Victor Markowitz, and Jean Thierry-Mieg we plan to reconsider the IGD prototype, and start building the whole integrated system from scratch, using the tools and concepts that have proven useful so far. It is not by chance that many of them are reported in this book.

There are six dimensions of integration we plan to follow:

1. Integration of data. We will design a comprehensive database for genome related conceptual and experimental objects, and populate it with data from major public databases and experimental resources.
2. Integration of the object and process information. Processes, such as laboratory protocols and data analyses, will be modeled together with 'static' data objects. This will enable to trace back the history of experimental objects or analysis results, and to use IGD locally as a laboratory notebook.

3. Integration of data and knowledge. We will develop a PROLOG interface to the IGD database to faciliate logic based analysis of the data, and the representation of biological knowledge, as well as of complex data integrity rules.
4. Integration of interfaces to analytical tools. We will develop an interface between the integrated database and major external software packages.
5. Visual integration of complex data and operations. The system will provide graphical display for complex objects like chromosome genetic and physical maps, clone grids, sequence feature maps, etc.
6. Possible integration of future databases and tools. The IGD system is designed as open and extensible. We will provide tools for evolutionary extensions of the database schema, for rapid data import from new databases, for the development of interface to new analysis programs, and for the implementation of new display methods.

6. ACKNOWLEDGMENTS

I wish to thank many people for their contribution to the IGD design, especially in its early phases: Sandor Suhai of DKFZ, Hans Lehrach and Steve Bryant of ICRF, Martin Bishop and Francis Rysavy of the UK HGMP-RC, and Shaun Cutts (now at The Brown University). Petr Kocab, Martin Senger, and Detlef Wolf of DKFZ have done great jobs in the implementation of the whole system. The implemented prototype takes much of its power from excellent tools and programs which were 'borrowed' and reused, first of all the ACEDB written by Richard Durbin of MRC and Jean Thierry-Mieg of CNRS/INSERM, and the (ERDRAW, SDT, COQL) family of database tools of Victor Markowitz of LBL. And finally I want to thank Annemarie Poustka and Thomas Fink of DKFZ for their experimental data and helpful feedback.

7. REFERENCES

1. Ritter, O., et al. "The IGD Project Documentation; InternalTechnical Report Nr.2, DKFZ Heidelberg, 1992. Available from o.ritter@dkfz-heidelberg.de.
2. Frenkel, K.A. "The Human Genome Project and Informatics", Communications of the ACM, 34, 41-51, 1991.
3. Keen, G., et al. "Access to molecular biology databases", Mathl. Comput. Modelling, 16, 93-101, 1992.
4. Benson, D., Boguski, M.S., Lipman, D.J. and Ostell, J. "The National Center for Biotechnology Information.", Genomics 6, 389-391, 1990.
5. Johnston, W., Lewis, S., Markowitz, V.M., McCarthy, J., Olken, F.and Zorn, M. "The Chromosome Information System", Technical Report LBL-29675, Lawrence Berkeley Laboratory, Berkeley, CA 94720, May 1991.
6. Zehetner, G., and Lehrach, H. The Reference Library Database (RLDB) In "Genome News" (N. Spurr, Ed.), No. 12, pp. 49-51. Imperial Cancer Research Fund, London, 1992.
7. Slezak, T., Wagner, M., Yeh, M., Cantu III, R. and Branscomb, E. "Livermore's Pragmatic Approach to Integrated Mapping for Chromosome 19", in this book.
8. see the ACEDB chapter of R. Durbin in this book.
9. Cherry, J. M., Cartinhour, S. W. and Goodman, H. M. "AAtDB, An Arabidopsis thaliana Database". Plant Molecular Biology Reporter volume 10, number 4, pages 308-309, 409-410, (1992).
10. ISO 9075: "Information Processing Systems - Database Language SQL", 1988.
11. Sybase SQL Server documentation, Sybase, Inc., Emeryville, CA 94608, USA.
12. Szeto, E., and Markowitz, V.M. "ERDRAW 2.2 Reference Manual", Technical Report LBL-PUB-3084, Lawrence Berkeley Laboratory, May 1991.
13. Markowitz, V.M., and Fang W. "SDT 4.1. Reference Manual", Technical Report LBL-27843, Lawrence Berkeley Laboratory, May 1991.

14. Markowitz, V.M., and Shoshani A. "QUEST Design Document", Preliminary Draft 1.6, Lawrence Berkeley Laboratory, September 1991.

15. "EMBL Data Library, User Manual", Release 28, European Molecular Biology Laboratory, Heidelberg, September 1991.

16. Bairoch, A. "The SWISS-PROT Protein Sequence Data Bank", Nucleic Acids Res. 19, 2247-2249 (1991).

17. Pearson, P.L., Matheson, N.W., Flescher, D.C., and Robbins, R.J. "The GDB(TM) Human Genome Database Anno 1992", Department of Medicine and Welch Medical Library, Johns Hopkins Univerisity School of Medicine, Baltimore, MD 21205-2100, USA, 1992.

18. McKusick, V.A., Francomano, C.A. and Antonarakis, S.E. "Mendelian inheritance in man: catalogs of autosomal dominant, autosomal recessive and X-linked phenotypes", Johns Hopkins University Press, Baltimore, USA, 1992.

19. Bryant, S. "UK DNA Probe Bank", System Specification Rev. 1.0, Imperial Cancer Research Fund, February 1991.

20. Lehrach, H., Drmanac, R., Hoheisel, J., Larin, Z., Lennon, G., Monaco, A.P., Nizetic, D., Zehetner, G., and Poustka, A. "Hybridization Fingerprinting in Genome Mapping and Sequencing", Genome Analysis Volume I: Genetic and Physical Mapping, Cold Spring Habor Laboratory Press 0-87969-358-4/90, 1990.

GENETIC MAPPING, AN OVERVIEW

Jacques S. Beckmann

Centre d'Etude du Polymorphisme Humain
27 Rue Juliette Dodu
Paris 75010, France

INTRODUCTION

Genetic analysis is likely to play an increasing role in the elucidation of the genomic architecture and their evolution, but also of the functions encoded in the genes themselves. It is but a matter of time before the entire sequence of each one of the 24 human chromosomes (22 autosomes and a pair of sex chromosomes) will be completely determined, i.e., the exact linear ordering of over 3 billion nucleotides constituting the human genome will be known. One may, therefore, wonder whether the interest in genetic studies will survive this fantastic challenge, the deciphering of the genome. It is our contention that this interest is genuine and will legitimately preoccupy scientists for years to come. The main reasons for this are that the message(s) of the DNA sequence information itself remains obscure, that we still lack the capacity to understand its many superimposed meanings. Adequate genetic questions will enable us to progress towards this endeavour. But there is also a second simpler reason for continuing with genetic studies. Indeed, not all genetic disorders have been characterised so far. And of those identified, almost all of the pathologies of complex etiology, such as multifactorial diseases, still await elucidation. Furthermore, our interest in biology goes beyond genetic disorders, and there will be a time where other traits will be in the limelight. For all these, genetics seems to be an inevitable path.

Under the term "genetics", we refer to the entire process that starts with the examination of a trait - the phenotype, and eventually leads to the identification and cloning of the underlying genetic structure(s) - the gene(s) - controlling it, and in this way contributes to our understanding of the physiological and underlying biochemical mechanisms. This is usually achieved by means of procedures generally referred to as reverse genetics or positional cloning [1, 2]. The identification of a gene whose biochemical defect is unknown starts first by its chromosomal assignment, by positioning it on the map - "the genetic map". For instance, a gene involved in the etiology of limb girdle muscular dystrophy was mapped to chromosome 15 region estimated to span about 20 million base pairs (megabases; [3]). But this assignment is way too large, in molecular terms, to allow to track and clone the gene. The gene's mapping co-ordinates need to be refined, as they will be used to launch the chromosomal walking that will lead to the establishment of a physical map of the region and eventually to the cloning of the gene. The current strategy is to narrow the genetic interval in which the target locus is located to its maximum before venturing into what can be an extremely long and arduous task, the cloning of the gene itself. But narrowing this interval implies that one can identify among the markers bracketing this gene those which are the closest. How this can be done will be briefly reviewed in this chapter. In so doing we shall refer mostly to examples or problems taken from human genetics, but these principles are general and can be applied to other studies as well.

PRINCIPLES AND DEFINITIONS

We shall start with a brief introduction of a few genetics terms. Almost all of the definitions given here are paraphrased from those found in Jurg Ott's excellent book [4]. Genetics deals, thus, with the study of heritable characters; these are determined by *genes*, the *phenotype* being the expression of a particular *genotype*. *Penetrance* is the conditional probability of observing the corresponding phenotype given the specified genotype. One estimates that there are some 50-100,000 genes in the human genome. And as we all have 2 copies (one paternal and one maternal) of each chromosome, each individual carries *two* copies of each gene. Each one of these may occur in different states, the *alleles*. Following *meiosis*, gametes contain only a *haploid* set of chromosomes. Each parent passes therefore to each of his/her children's *one of two* alleles with probability 1/2. Genetic studies are mostly done on families such as that depicted in Figure 1, and the reader may find it helpful to accompany this reading with an examination of the segregation of the different marker alleles depicted in this Figure.

Let us consider now two genes simultaneously, say M1 and M2 in Fig. 1. The parental allelic distribution of the markers, or *"phase"*, is determined by the grand-parental genotypes. These are not always known, but one can also see from Figure 1 that it can be inferred from the children's genotypes, provided they are numerous. One defines a *haplotype* as the combination of alleles (at different genes) received by an individual from one parent. There are thus two types of grandparental haplotypes among the children: parental and non-parental, i.e. *recombinant*. In other words a recombinant haplotype will have alleles from two grandparents.

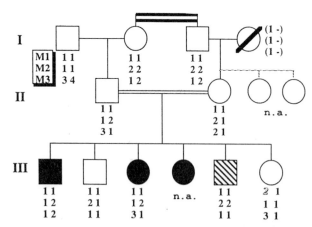

Figure 1. Schematic representation of a three (I,II, and III) generations family. M1, M2 and M3 are three different genes for which the respective alleles are given as numbers. Square and round symbols represent male and female individuals, respectively. Other symbols are as follows: diagonal slash: deceased person; open and filled symbols: healthy and affected members, respectively; hatched symbol: uncertain phenotype; double horizontal line: consanguineous marriage, indicating that two of the grandparents are related; numbers in parenthesis: inferred genotypes; n.a: not available.

Let us consider alleles of genes on different chromosomes (say genes M2 and M3 in Fig. 1). Their respective alleles will segregate independently of each other at meiosis, i.e., they will have an equal *a priori* probability (of 25%) of being transmitted together to the same gamete. Contrast this with syntenic genes (i.e., on the same chromosome; say M1 and M2), these will tend to be transmitted as a "linked" pair, depending on whether or not they recombined, i.e., were separated by a chromosomal crossing over event. There will thus be an excess of parental versus non-parental types for closely linked markers. The further apart the genes are on a chromosome, the more they will recombine (and eventually behave like non-syntenic genes). To sum up, for genes which are on different chromosomes the alleles from a specific grandparent will be coinherited 50% of the time, and there will be 50% recombinants, whereas for linked genes we will observe a lower frequency of recombination (between 0 and 50%). Thus, the rate of crossovers between two genes can serve as a stochastic measure of the distance between them. It should be remembered, however, that crossovers are not directly observable. Recombinations are (Figure 2). One can thus measure recombination frequencies (theta). But this parameter, as it is bound by the upper 50% value (see ref. [4] for a more detailed justification), is not a linear reflection of the genetic distance. It is thus not additive, except for its lowest values. Fortunately, there exists a number of different empirical formulas which enable the conversion of % recombination (θ) into an additive genetic distance, the centiMorgan (cM). To give a sense of perspective you should know that the human genome covers over 3000 cM, with one cM averaging about one megabase of DNA.

OF TRAITS, FAMILIES AND MARKERS

Measuring recombination events, implies that linkage analyses be performed. In other words, in order to study the segregation of **characters** in **families, polymorphic loci** (markers, e.g. RFLPs, ...) need to be available. The purpose is thus, to look for possible co-segregation of particular markers with the phenotype of interest. The marker must, by definition, be polymorphic as it will be used to monitor the genetic transmission of the corresponding chromosomal segment throughout the families. If it is not polymorphic, it will be said to be non-informative in that particular family (such as M1 in the family depicted in Fig1, or M2 for the grandparents). Informative families are thus those in which at least one of the parents is a double heterozygote at both loci examined.

Until not too long ago, markers were few and hard to find. Fortunately, since 1980, it is possible to use several means to uncover extremely rich sources of genetic variability. It is not our purpose to review these markers and the procedures used to detect them. Suffice it to say, that they all share the same property, namely, that they involve monitoring variability at the DNA level directly (quite a long way away from having to infer genotypes from a phenotypic observation). The most promising of all these different types of genomic markers seem to be the group commonly referred to as micro satellites [5, 6], of which the (CA)- based repeats are the most prominent representatives. A human genetic map consisting of 813 of these markers was recently reported [7]. Similar progress was also made in the

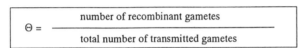

$$\Theta = \frac{\text{number of recombinant gametes}}{\text{total number of transmitted gametes}}$$

The recombination rate can be converted in genetic distance

===> Genetic distance: expressed in centiMorgans (cM)
1 cM = ± 1 % recombination

Figure 2. Frequency of recombination between two loci

development of micro satellite-based genetic maps for the mouse [8] and rat [9] genomes. Use of these markers, as a result of the large number of alleles encountered at each locus, will tend to maximise the probability that the families will be informative.

Markers permit us to identify, whenever these exist, genetic linkages between loci. Markers (i.e., specific marker alleles) will be scored in families in which the traits of interest are segregating, possible recombinants identified (In the pedigree shown in Fig.1, the first son is a recombinant between M2 and M3, having inherited a partial haplotype from two different grandparents, and he is not the only one!), and the frequency of meiotic recombinants will be determined. The number of scored meioses must, however, be sufficient for this value to reach statistical significance.

This point leads us straight to the requirement for an adequate number of families segregating for the trait of interest. Assembling these pedigrees may not always be an easy task. Human families are seldomly comprised of large sib ships. One has to take them as they come, even when specific individuals refuse to take part in the study or are simply missing, i.e., unavailable for the analysis. Some diseases may be infrequent. There are other problems too (Table 1). What appears to be a single "well" defined phenotype may hide a genetically heterogeneous disease. In other words, mutations in different genes may have similar phenotypic expressions and may result in the fusion of different genetic entities into one artificial nosological group. Some traits may be under the control of a number of genetic and non-genetic factors; they may have a low heritability and be under strong environmental influences. All these and other factors listed in Table 1 contribute to complicate the genetic analysis.

In selected instances there may be an alternative. It consists of examining an animal model population which has a phenotype that is as close as possible to the phenotype of interest, and to map the genes involved. One can then resort to comparative mapping (for a discussion of these methods see reference [10]) to define the homologous human chromosomal segments and test their potential role in the determination of this phenotype in humans. Valuable experimental populations have been developed, e.g. diabetic or hypertensive strains of mouse or rat. The relative advantages of animal models are briefly summarised in Table 1. Such studies have already led to important initial successes (e.g. references [11, 12]). There are, however, critical limitations. It is indeed not granted that the genes involved in a particular phenotype in the model, are playing a analogous role in humans. The ACE (angiotensin converting enzyme) story is a good example. The rat studies although they never really incriminated ACE itself, nevertheless suggested that this gene might play an important role in regulating hypertension not only in the rat, but also in man. Subsequently, it was shown that ACE is most likely not involved in these processes in man [13], but rather has a role in predisposing to myocardial infarction [14]. It is thus crucial that the deduction based on the examination of animal models be validated by human studies.

THE DETERMINATION OF GENETIC LINKAGE

There are different genetic approaches to identify linkage between loci. Only two will be considered here in more detail. Both rely on the study of the joint transmission, or co-segregation, of the two markers, or the phenotypic trait, say a disease, and the genetic marker in affected families. In the first case, in the absence of any *a priori* mapping knowledge, a panel of markers representative of the entire genome will be used to systematically scan for linkage in one chromosomal segment after another. This is often referred to as the random marker approach and involves testing some 200 markers for the entire genome. When previous mapping information is available confining the gene to a particular region, the analysis can be restricted to markers mapped into this area. The second approach is a specific case of the first one. Indeed, now and then, a gene has been identified whose function suggests that this particular gene may be involved in the etiology of the disease studied. It can therefore be con-

Table 1. Relative advantages of animal models versus human populations

Human studies

o   ascertainment of informative families
o   size and incompleteness of families
o   consanguineous marriages
o   number of meioses can be limiting
o   complex set of genetic causes within and between families
o   clinical and genetic heterogeneity
o   imprecise phenotypic definition
o   phenotype modulated by environmental influences
o   low heritability of trait

Experimental animal models

o   single set of genetic causes within each inbred strain
o   controlled test crosses with divergent parents
o   repeatability of cross
o   controlled number of (phase known) meioses
o   availability of special genetic lines (e.g. recombinant inbred or congenic strains)
o   control of experimental environment
o   higher heritability
o   comparative mapping
o   invasive phenotypical (physiological) characterisation
o   availability of tissues for biochemical/physiological studies
o   developmental studies
o   generation of transgenic animals
o   knock out of genes
o   generation of new mutations
o   complementation of mutant phenotypes

But homologies with human genes are not straightforward

o   ... must demonstrate that the homologous gene functions in same physiological pathway
o   ... relationship to human disorder must be established ...

---

sidered as a candidate gene. If one possesses a marker for this gene, the cosegregation of candidate marker and disease can readily be tested. This candidate gene approach is very powerful and is expected to become more and more prominent as more genes are placed on the map. It is possible that the gene has been only vaguely localised, or that there are no suitable genetic markers for it, or that comparative mapping inferences from mouse or another experimental model suggest that in humans a particular homologous region may contain such a gene. One then talks about a candidate region. Testing for linkage of hypertension with ACE in humans is one example of this candidate gene approach. Another very successful story is the demonstration that a battery of different mutations in the glucokinase (GCK) locus are each involved in the generation of a juvenile form of insulin-independent diabetes, known as MODY [15, 16].

Given that the marker information for all families has been collected, how does one proceed? One calculates the likelihood for observing this marker segregation under two different assumptions, that the two loci considered are linked, H1 (i.e., $0.0 =< \theta < 0.5$) or in the null hypothesis, H0, unlinked ($\theta=0.5$). Consider the family depicted in Figure 3. There are five

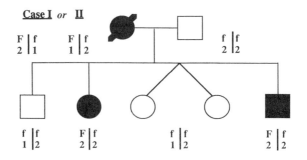

lod score: $\log_{10}[L(\theta)/L(\theta=0.5)]$

theta:	0.0	0.01	0.05	0.10
lod score:	0.903	0.886	0.814	0.720

Figure 3. Segregation of a dominant disease phenotype (F) in a hypothetical pedigree with 5 children of which 2 healthy daughters (f,f) are monozygotic twins. The alleles at a second locus are given as numbers. Vertical bars separating the alleles denote that the phase is known.

children, 2 of which are monozygotic twins. As the latter reflect the same meiotic events, only 4 children will be considered in the genetic analysis. Notice also that the mother is given as deceased. We can infer her genotype, but not the phase. Let's first assume that case I can be demonstrated. In this instance all 4 children will be of the non-recombinant type. The likelihood for this family is simply $L(\theta) = (1-\theta)^4$. Were case II to represent the mothers' phase, all children will be recombinants, and hence the family likelihood will be $L(\theta) = \theta^4$. As in reality phase is unknown, and since there is an equal *a priori* probability of encountering either phase, the family likelihood becomes: $L(\theta) = 1/2[(1-\theta)^4 + \theta^4]$. It is easy to derive the likelihood estimate for this family as a function of theta. Following a suggestion made in reference [17] one defines the lod score (logarithm of the odds) as the base 10 log of the likelihood ratios $[L(\theta<0.5)/L(\theta=0.5)]$. Figure 3 shows the values that can be derived for this family. qmax indicates the most probable distance separating two markers, and it is that value of theta at which the maximum lod score (Zmax) is obtained.

In the example given, it is rather easy for this small family to derive the likelihood for different values of theta. All these calculations, however, quickly become impossible without a computer's help as the families get larger and more complex and as the number of missing individuals increases. We also considered in this example that the recombination rates in male and female gametes were identical. This, however, is not the case and there is no uniform correspondence between these, as the ratios of male to female genetic distance can vary from one chromosomal region to another. One may want to derive maximum likelihood estimates under the model of sex-specific recombination rates. Fortunately there exist a number of software packages, such as LIPED [18], LINKAGE [19]), MAPMAKER [20], CRIMAP [21], MAP [22] and MENDEL[23], that enable one to determine lod score values and/or generate multipoint maps. The lod score test has many important advantages. It enables one to sum over all informative families without considering the number of generations, the specific study, etc. This is shown in Table 2, which reproduces part of the results obtained in the MODY study following the candidate gene approach [15, 16]: each family contributes its share to the lod score value, which is added onto that of the other families. Table 2 also illustrates how this study led to the identification of a number of different mutations in this gene, for which lod score values are given, as well as a series on neutral and unlinked polymorphism.

Table 2. Glucokinase Mutations (partial list)

Type	Location	Family	a.a. change	Lod scores (θ=0.0)
Neutral Variant Sequence Site	Exon 1a	F370	Ala-->Ala	
		F8	Asp-->Asn	
	Exon 3	F51	Thr-->Thr	
Missense	Exon 5	F386	Gly-->Arg	0.52
		F114	Val-->Met	0.23
	Exon 6	F422	Val-->Ala	1.52
	Exon 7	F391	Thr-->Met	1.15
		F388	Gly-->Arg	1.40
		F390	Gly-->Arg	0.30
	Exon 8	F28	Glu-->Lys	2.22
		F51	Glu-->Gln	14.10
		F160	Leu-->Pro	1.30
Splice site	Exons 6/7	F85		1.74
	Exons 9/10	F397		0.55
Total lod score				**25.03**

In practice, if Zmax for a given value of theta is greater than or equal to 3, by convention the loci are considered to be in linkage with a significance level of 0.001. Linkage can be excluded in the region for which $Z(\theta)$ is smaller than -2 [the null hypothesis (absence of linkage) is accepted with a significance level of 0.01]. If -2 > Zmax > 3, the data are inconclusive. More informative meioses need to be examined (more meioses or more informative markers).

So far we considered the cosegregation of two loci at a time, the traditional two-point comparison. One can also study a larger number of loci simultaneously. This implies deriving likelihood functions over all markers and thetas. This allows ordering the loci with respect to each other, and hence, to generate multipoint maps. Currently all maps are computed by considering multiple loci using one of the programs mentioned above.

To illustrate the power of these methods, we will recall the rapid progress that was made in establishing a reference map for the human genome. This is the result of an ongoing international collaboration wherein close to 100 laboratories all contribute genotype information on the same reference families, known as the CEPH Reference Panel [24, 25]. This collaboration has already led to the assignment on the human genetic map of close to 4000 genomic markers, of which over 25 % are highly informative micro satellites. Current efforts are now geared towards the generation of high quality maps with a resolution within the 1 cM range. Another way to illustrate the rate at which genetic mapping progresses, is by looking at the increasing list of genetic diseases that were mapped since 1983. Table 3 lists just a few of examples.

Without being exhaustive, one should mention that alternative non-parametric methods have also been developed (for more references see ref. [4]). Indeed, traditional linkage approaches like those described above require that specific assumptions be made as to the mode of inheritance, gene frequencies, and other parameters. These, however, are not always easy to drive. Non-parametric methods such as the affected sib-pair analysis are based only on information from affected individuals. They are thus not sensitive to an inadequate estimation of penetrances. The rationale of these methods is that affected sibs will more frequently share the same parental alleles at marker loci closely linked to the gene of interest than at unlinked loci. Association studies done at the population level are also very useful. Both types of methods are likely to be of great interest in the study of complex multifactorial characters.

Table 3

Diseases localised	chromosomes	genes cloned
Duchenne muscular dystrophy	Xp21.3-p21.1	DMD
Cystic fibrosis	7q31-q32	CFTR
Retinoblastoma	13q14.2	RB1
Neurofibromatosis	17q11.2	NF1
Familial Polyposis	5q21-q22	APC
X-linked mental retardation	Xq27.3	FMR-1
Cardiac hypertrophy (form1)	14	MYO
Maturity onset diabetes of the young	7p	GCK
Limb girdle muscular dystrophy 1	5q	
Limb girdle muscular dystrophy 2	15q	
Huntington Chorea	4p14.3	
Spinocerebellar ataxia	6p24-p21.3	

NEW DEVELOPMENTS, NEW NEEDS

A word of caution is needed here. Linkage assignments are essentially statistical determinations, which can be considered as the end product of a long multi-disciplinary succession of steps. These will start with the definition of the phenotype of interest, followed by the assembly of informative families, the determination of the inheritance mode for the trait of interest, and the generation of the marker genotypes. As a result these assignments are extremely vulnerable: different types of errors (examples can be seen in Figure 1 and in Table 4) can occur in this long series of steps, some of which could be fatal. It is thus essential that great care be taken throughout this work to minimise such errors. Figure 1, for instance, illustrates a pedigree showing a consanguineous relation, where a number of members were not available for the study, and where the healthy daughter had at the first marker locus (M1) an allele which was not seen in either parent. The latter could be due to a genotyping or other sampling error. Alternatively, the appearance of this allele could be the result of a neomutation, or simply a case of erroneous paternity assignment.

One last word, no discussion on mapping would be complete without a reference to other mapping progress. One should indeed emphasise that following the achievements realised in genetic mapping, the effort towards a complete physical map of the human genome has made real progress in 1992. This has been realised by the elaboration of a single contig, i.e., an uninterrupted contiguous series of overlapping YACs (yeast artificial chromosomes), covering the entire length of the long arm of chromosome 21 [26] or the chromosome Y [27] A second example of a similar success story is the attempt to group human YACs in contigs on the basis of overlapping fingerprints [28]. It is clear that as these physical mapping efforts progress, these will complement the genetic work. Information on order will be derived faster from physical mapping data. New markers will be deliberately derived for a particular chromosomal region in a targeted fashion. And last, but not least, information on the physical map is expected to play a central role in positional cloning.

In the proceeding, we tried to give a quick review of the principles, methods and caveats of genetic mapping approaches. This was accompanied by a couple of examples. Clearly, geneticists have addressed the most frequent and simplest mapping problems first.

These involved problems wherein alternative phenotypic classes were easy to recognise as they fell into discrete distributions. With time, the interest focuses more and more on so called complex multifactorial problems. This leads to new difficulties which are often further

Table 4. Vulnerable areas of the linkage method

Sources of errors:

o   Relaxed clinical ascertainment criteria
o   Inadequate diagnosis (misdiagnosis, misclassification)
o   Erroneous P(M)aternity assignment
o   Sample mix-ups
o   Incomplete digestions (Southern blots)
o   Incorrect interpretation of the genotypes
o   Inappropriate genetic models
     (mode of inheritance, penetrance, phenocopies, allele freq., ...)
o   Data entry errors

Potential complications:

o   Cryptic genetic heterogeneity
o   Germline mosaicism
o   Uniparental disomy
o   New mutations

Consequences of these errors:

o   Appearance of multiple recombinants
o   Omission of authentic recombinants
o   Increase in genetic distances
o   Selection of incorrect map orders
o   Erroneous exclusion inferences

<u>**Dramatic for:**</u>   o **genetic analyses**
                  o **positional cloning**

compounded by genetic heterogeneity. Addressing these problems will require new, faster, more powerful algorithms than are presently available. These should enable the treatment of continuous, often overlapping or skewed distributions. They should allow that non-genetic factors be taken into consideration (such as socio-cultural or environmental ones). Further-more, while the introduction of multiallelic markers is seen as a boon, it also creates new computing problems. A list of desired options could be made. It was only touched here to stress that geneticists are in urgent need of new computational tools.

ACKNOWLEDGMENTS

I would like to express my warmest thanks to Drs. H. Cann, D. Weeks, D. Grausz, V. Teichberg and Ph. Froguel for their help in writing this manuscript.

REFERENCES

1. Collins, F. 1992. Positional cloning: Let's not call it reverse anymore. Nature genetics 1: 3-6.
2. Orkin, S.H. 1986. Reverse genetics and human diseases. Cell 47: 845-50.
3. Beckmann, J.S, I. Richard, et al., (1991). A gene for limb-girdle muscular dystrophy maps to chromosome 15 by linkage. C. R. Acad. Sci. 312, Série III: 141-148.
4. Ott, J. 1991. Analysis of Human Genetic Linkage. Baltimore: John Hopkins University Press.

5. Litt, M., and J. A. Luty. 1989. A hypervariable microsatellite revealed by *in vitro* amplification of a dinucleotide repeat within the cardiac muscle actin gene. Am. J. Hum. Genet. 44: 397-401.

6. Weber, J. L., and P. E. May. 1989. Abundant class of human DNA polymorphisms which can be typed using the polymerase chain reaction. Am. J. Hum. Genet. 44: 388-396.

7. Weissenbach, J., G. Gyapay, C. Dib, A. Vignal, J. Morissette, Ph. Millasseau, G. Vaysseix and M. Lathrop. 1992. A second generation linkage map of the human genome. Nature 359: 794-801.

8. Dietrich, W., H. Katz, S.E. Lincoln, H.-S. Shin, J. Friedmann, N.C. Dracopoli and E. S. Lander. 1992. A genetic map of the mouse suitable for typing intraspecific crosses. Genetics 131: 423-47.

9. Serikawa, T., T. Kuramoto, P. Hilbert, M. Mori, J. Yamada, C.J. Dubay, J.-L. Guenet, G.M. Lathrop, and J.S. Beckmann. 1992. Rat gene mapping using PCR-analyzed microsatellites. Genetics 131: 701-721.

10. O'Brien, S.J., J.E. Womack, L.A. Lyons, K.J. Moore, N.A. Jenkins and N.G. Copeland. 1993. Anchored reference loci for comparative genome mapping in mammals. Nature genetics 3: 103-112.

11. Hilbert, P., K. Lindpaintner, J. S. Beckmann, T. Serikawa, F. Soubrier, C. Dubay, P. Cartwright, B. De Gouyon, C. Julier, S. Takahasi, M. Vincent, D. Ganten, M. Georges and G. M. Lathrop, 1991 Chromosomal mapping of two genetic loci associated with blood-pressure regulation in hereditary hypertensive rats. Nature 353: 521-529.

12. Jacobs, H. J., K. Lindpaintner, S. E. Lincoln, K. Kusumi, R. K. Bunker, Y. -P. Mao, D. Ganten, V. J. Dzau and E. S. Lander. 1991. Genetic mapping of a gene causing hypertension in the stroke-prone spontaneously hypertensive rat. Cell 67: 213-224.

13. Jeunemaitre, X., R.P. Lifton, S.C. Hunt, R.R. Williams and J.-M. Lalouel, 1992. Absence of linkage between the angiotensin converting enzyme locus and human essential hypertension. Nature genetics 1: 72-75.

14. Cambien, F., O. Poirier, et al., 1992. Deletion polymorphism in the gene for angiotensin-converting enzyme is a potent risk factor for myocardial infarction. Nature 359: 641-4.

15. Froguel, Ph., M. Vaxillaire, F. Sun, G. Velho, H. Zouali, M.O. Butel, S. Lesage, N. Vionnet, K. Clément, F. Fougerousse, Y. Tanizawa, J. Weissenbach, J.S. Beckmann, G.M. Lathrop, Ph. Passa, M.A. Permutt and D. Cohen. 1992. The glucokinase locus on chromosome 7p is closely linked to early onset non-insulin dependent diabetes mellitus. Nature 356: 162-164.

16. Froguel, P., H. Zouali, N. Vionnet, G. Velho, M. Vaxillaire, F. Sun, S. Lesage, M.-O. Butel, M. Stoffel, J. Takeda, P. Passa, A. Permutt, J.S. Beckmann, G. Bell and D. Cohen. 1993. Familial hyperglycemia due to mutations in glucokinase: Definition of a subtype of diabetes mellitus. N. Engl. J. Med. 328: 697-702.

17. Morton. N.E. 1955. Sequential tests for the detection of linkage. Am. J. Hum. Genet. 7: 277-318.

18. Ott, J. 1974. Estimation of the recombination fraction in human pedigrees: Efficient computation of the likelihood for human linkage studies. Am. J. Hum. Genet. 26: 588-97.

19. Lathrop, G. M., and J. M. Lalouel, 1984 Easy calculations of lod scores and genetic risks on small computers. Am. J. Hum. Genet. 36: 460-465.

20. Lander, E.S., P. Green, J. Abrahamson, A. Barlow, M.J. Daly, S.E. Lincoln and L. Newburg. 1987. MAPMAKER: An interactive computer package for constructing primary genetic linkage maps of experimental and natural populations. Genomics 1: 174-81.

21. Goldgar, D.E.P., P. Green, D.M. Parry and J.J. Mulvihill. 1989. Multipoint linkage analysis in neurofibromatosis type I: An international collaboration. Am. J. Hum. Genet. 44: 6-12.

22. Morton, N.E. and V. Andrews. 1989. MAP, an expert system for multiple pairwise linkage analysis. Ann. Hum. Genet. 53: 263-69.

23. Lange, K., D. Weeks and M. Boehnke. 1988. Programs for pedigree analysis: MENDEL, FISHER, and dGENE. Genet. Epidemiol. 5: 471-72.

24. Dausset, J., H. Cann, D. Cohen, G.M. Lathrop, J.-M. Lalouel and R. White. 1990. Centre d'Etude du Polymorphisme Humain (CEPH): Collaborative genetic mapping of the human genome. Genomics 6: 575-77.

25. NIH/CEPH Collaborative Mapping Group. 1992. Science 258: 67-86.

26. Chumakov, I., P. Rigault, et al., 1992. Continuum of overlapping clones spanning the entire human chromosome 21q. Nature 359: 380-7.

27. Foote, S., D. Vollrath, A. Hilton and D.C. Page. 1992. The human Y chromosome: Overlapping DNA clones spanning the Euchromatic region. Science 258: 60-66.

28. Bellanné-Chantelot, C., B. Lacroix et al., 1992. Mapping the whole human genome by fingerprinting yeast artificial chromosomes. Cell 70: 1059-1068.

REPRESENTING GENOMIC MAPS IN A RELATIONAL DATABASE

Robert J. Robbins

Applied Research Laboratory
Johns Hopkins University
2024 East Monument Street
Baltimore, MD 21205 USA

INTRODUCTION

The goals of the Human Genome Project are: (1) construction of a high-resolution genetic map of the human genome, (2) production of a variety of physical maps of all human chromosomes and of the DNA of selected model organisms, (3) determination of the complete sequence of human DNA and of the DNA of selected model organisms, (4) development of capabilities for collecting, storing, distributing, and analyzing the data produced, and (5) creation of appropriate technologies necessary to achieve these objectives.

Given the amount of data that will be generated as progress toward these goals is made, it is imperative that electronic means for storing and manipulating the data be available. Databases must be built to describe map objects and mapping reagents, and to accommodate genetic and physical genomic maps as they are produced.

As our understanding of the human genome grows, the concepts that must be represented in these databases will increase in complexity and subtlety. Since these databases are expected to become a new scientific literature through which electronic data publishing will occur [1, 2, 3], they must be designed to handle the changing concepts of "gene" and "genomic map" without requiring major redesign each time a new finding occurs. The data models used must be sufficiently complex and abstract to represent all of our present concepts of genes and maps, as well as to evolve gracefully with the findings on genomic anatomy.

Although the word "gene" may be the most frequently used word in biology, it has proven remarkably difficult to define. Entire books have been written describing the early history of the gene concept [4], and many eminent biologists addressed the question during the classical period of genetics [5, 6, 7, 8]. In the modern era, major textbooks on molecular and cellular biology all devote significant efforts to defining the gene (e.g. ref. [9, 10], with one recent work [11] simply claiming that no single definition of the gene exists :

> The unexpected features of eukaryotic genes have stimulated discussion about how a gene ... should be defined. Several different possible definitions are plausible, but no single one is entirely satisfactory or appropriate for every gene.

If a gene cannot be defined, then how is one to design a data model to represent them? And, without a definition for genes, how possibly could we represent genomic maps? It is a truism in information science that an adequate data model cannot be developed without an understanding of the thing being modelled. Therefore, to build good databases we must turn our attention to the notion of the gene.

WHAT IS A GENE?

The gene, originally described as the hypothetical fundamental unit of heredity, is now known to consist of instructions encoded in the nucleotide sequence of a DNA molecule. To see why advances in our understanding of gene structure have complicated our notion of what a gene is, let us consider briefly the history of the gene concept.

Insights from Classical Genetics

The first notion of the gene came from Mendel's breeding studies on peas, in which he showed that patterns of inheritance were consistent with the assumption that the traits of each individual were controlled by a pair of independent factors, with one received from each parent. After 1900, Mendel's findings were rediscovered and extended, especially by the *Drosophila* group at Columbia under T. H. Morgan.

Very early, Morgan hypothesized and Sturtevant showed (Figure 1) that the inheritance patterns associated with genes believed to be carried in the same linkage group could be explained by assuming that linkage groups were chromosomes and that genes were carried on chromosomes in a regular, linear order.

When recombinational mapping proved generally applicable to all organisms, the resulting classical concept held that genes are (1) the fundamental unit of heredity, (2) subject to rare mutation, (3) stable across generations, (4) carried on chromosomes, and (5) capable of recombining during meiosis. According to these properties, two genetic traits that could not be separated by recombination were thought to involve mutations to the same gene, whereas traits that could be separated were held to involve two different genes.

The view of genes as essentially indivisible fundamental units occupying fixed chromosomal positions was summarized by Sturtevant and Beadle [13]:

> The relative constancy of crossover values and the constant order of genes in chromosomes imply that every gene occupies a fixed position in a chromosome, and its allele a corresponding position in a homologous chromosome. ... Such a position is known as a *locus*. ... [By asserting] the linear arrangement of genes in chromosomes, we do not, of course, imply by linear arrangement a straight line, but rather that the genes are arranged in a manner similar to beads strung on a loose string.

This "beads on a string" metaphor, so characteristic to the classical view of the gene, carried with it several corollary notions that have provided impediments to clear thinking about gene mapping, even to the present day. These include the idea of the indivisible gene and of the existence of a "string" -- a scaffolding molecule that could provide a coordinate space on which genes might be placed independently of the presence or absence of other genes. The coordinate-space problem will be discussed in more detail later.

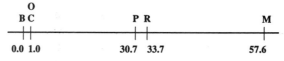

Figure 1. The first genomic map, as derived by Sturtevant [12] from recombination data in Drosophila. "B" = yellow body, "C" = white eye, "O" = eosin eye, "P" = vermilion eye, "R" = rudimentary wing, "M" = miniature wing. The distances are given as percent recombinants.

Modelling the Classical Gene in a Relational Database

The classical model of the genome lends itself to a very straightforward data model for representing genes and maps. Genes become nodes and pair-wise orderings and distances become arcs in a directed acyclic graph (DAG). DAGs are fine data structures for representing partial ordering among defined objects, and DAGs are relatively easy to implement in a relational database. One or more entity tables can be designed so that each tuple contains the information describing an individual gene or other map object (a node). Tuples in a separate relationship table can represent arcs, by containing the identifiers denoting two map objects joined by an arc. Additional attributes of the arc (like measured distances) are easily added to the relationship table. Although no standard SQL commands currently exist for executing the necessary transitive closure queries over the DAG, methods for implementing such queries are well known.

The Discovery of Pseudoalleles Challenged the Classical View

The classical notion of the gene was shaken when the first instance of apparent intragenic recombination was observed. The process was said to define "pseudoalleles," since by definition, true alleles could not recombine [14]. In 1955, Bentley Glass [15] wrote, "Fifty years from now it seems very likely that the most significant development of genetics in the current decade (1945-1955) will stand out as being the discovery of pseudoallelism." Although this claim now seems wildly inaccurate (especially since that is the same decade in which the structure of DNA was first established and bacterial genetics was founded), the strength of the sentiment shows just how firmly held was the notion of the indivisible gene.

The interest in pseudoalleles proved short-lived, as the DNA-sequence concept of the gene made intragenic recombination seem inescapable, not implausible. When Benzer's development of the cis-trans complementation test rendered obsolete the old recombinational test for gene individuality [16], the storm over pseudoalleles faded away. The complementation test establishes whether two genetic traits involve the same or different genes by testing whether or not two chromosomes, each carrying a different defective gene, complement each other's deficit and restore normal function. If complementation occurs, the defects are held to occur in different functional units. If not, the defects are presumed to occur in the same functional unit. Benzer coined the term "cistron" to describe the functional units so identified.

The concept of the cis-trans test as defining genetic functional units became so widespread that many began to equate the cistron with the gene, as in the following textbook definitions:

Cistron A nucleotide sequence in DNA specifying a single genetic function as defined by the complementation test; a nucleotide sequence coding for a single polypeptide; a gene.[7]

Cistron Originally defined as a functional genetic unit within which two mutations cannot complement. Now equated with the term gene, as the region of DNA that encodes a single polypeptide (or functional RNA molecule such as tRNA or rRNA). [18]

The Early Molecular Perspective

The molecular notion of the gene originated from biochemical studies, first on eye color in *Drosophila* and later on nutrient requirements in *Neurospora*, showing that individual genes seemed to be associated with the presence of individual functional enzymes. From this, the famous *one-gene, one-enzyme* hypothesis was proposed, then modified to *one-gene, one-polypeptide*.

With the discovery of the structure of DNA and the elucidation of the triplet code, the *one-gene, one-enzyme* concept was extended to *one-gene, one-macromolecule* and a gene became identified as that stretch of DNA responsible for determining a particular polypeptide or RNA. Such a definition is still commonly encountered today:

Gene A hereditary unit that, in the classical sense, occupies a specific position (locus) within the genome or chromosome; a unit that has one or more specific effects upon the phenotype of the organism; a unit that can mutate to various allelic forms; a unit that codes for a single protein or functional RNA molecule [19].

The early molecular model of the gene can be summarized in either an operational, functional definition (the cistron detected with a complementation test) or a structural *one- gene, one-product* definition (the DNA sequence encoding a functional macromolecule).

Modelling the Early Molecular Gene in a Relational Database

In either of these molecular definitions, the gene is no longer indivisible, but it is still discrete and contiguous. And, it still has a well defined unitary function -- the specification of its protein or RNA product. By extending the notion of the gene so that it has linear extent, the data model developed for the classical gene can easily be used to represent the early molecular concept of the gene.

Current Molecular Perspectives on the Gene

Continuing work on the regulation of gene expression began to show that regions of DNA not directly involved in specifying the sequence of a protein were nonetheless essential in determining its level of production. Although the discovery in the late 1970s that some genes contained introns upset the simplistic notion that genes and their proteins products were perfectly colinear, the early molecular model of the gene required only minor modifications to accommodate new findings so long as a continuous region of DNA produced one continuous polypeptide or RNA.

However, additional discoveries have begun to disturb the early molecular model of the gene just as surely as the discovery of pseudoalleles refuted the classical definition. Complex regulatory units, with very subtle regulation have been described. Some regulatory units, such as the SOS and heat-shock regulons in *E. coli*, were found to control the expression of non-contiguous genes.

The discovery of overlapping coding regions has undercut (and the finding of nested genes within the introns of other genes destroyed) the idea of the gene as a necessarily discrete unit. Even more complex structures, such as nested gene families call into question the definitiveness of the complementation test. The occurrence of complex post-translational processing (whereby one transcript yields one polypeptide, which is then processed into multiple different functional peptides) leads to counter-intuitive results when combined with some definitions of the gene. Most challenging of all, the discovery of RNA editing (where the transcript of one DNA segment is enzymatically modified, under the direction of an RNA transcript from another piece of DNA, to yield a functional mRNA) undermines the notion of the gene as a contiguous region of DNA.

Complex Regulation. The macromolecular synthesis (MMS) operon (Figure 2) in *E. coli* could also be called the "fundamental dogma" operon, since its three protein products are involved in DNA replication, transcription, and translation. Given the divergent times at which these processes occur, it is difficult to imagine how all three proteins could be effectively regulated in a single simple operon.

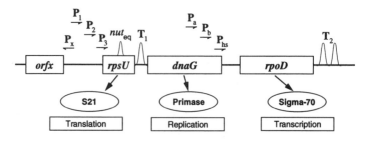

Figure 2. The complex macromolecular synthesis operon in *E. coli*, as described by Lupski and Godson [20]

However, in addition to normal operon control, the MMS operon contains a maze of complex, overlapping control mechanisms. The operon has six promoters (seven, if the P_x promoter for "orf$_x$," an open reading frame of unknown function is included). The "P_1," "P_2,"and "P_3," promoters control transcription initiation for the operon as a whole. Two other promoters, "P_a" and "P_b", also affect the *rpoD* locus, and these additional promoters are embedded in the coding region of the *dnaG* locus. Another promoter, "P_{hs}", is a heat-shock promoter. "T_2" indicates the main transcription termination signal, and "T_1" indicates an alternate terminator.

With all of these overlapping control and coding regions, the definition of precise boundaries and extents for the genes in this operon are subject to reasonable debate among competent biologists.

Alternate Splicing and Nested Genes. The discovery in the early 1980s that some regions of DNA can produce more than one polypeptide through the alternate splicing of its transcript product began to undermine the generality of the *one-gene, one-product* notion of the gene. The discovery of genes nested within the introns of other genes further eroded the concept of genes as continuous, discrete regions of DNA.

The *Gart/Lcp* loci in *Drosophila* (Figure 3) illustrate both of these situations. The transcript of the *Gart* locus can undergo alternate splicing to yield two different gene products, and the *Lcp* locus resides entirely within the large intron of the common region of the *Gart* splice options. Given that the two loci are encoded on opposite strands of the DNA, it is arguable whether it is better to consider these two loci as being nested, or whether it is better to consider the outer locus to be discontinuous. One might even contend that, in general, loci should be assigned to regions of a particular strand of DNA, in which case the *Gart/Lcp* locus is unexceptional.

Nested Gene Families. A recent study on the human UDP-glucuronosyltransferase locus on chromosome 2 has found a complex region, with six promoters, six duplicated (and diverged) alternate first exons, and four common trailing exons (Figure 4). Since each alternate first exon has its own promoter, one might suggest that this is not a case of alternate RNA splicing, but rather a case of a nested gene family, where only part of the gene has been duplicated, but where divergent evolution has occurred nonetheless.

This nested structure challenges the cistron notion of the gene, since mutations in different alternate first exons should complement, whereas mutations in the other four exons would not. Under either the cistron definition of the gene or the *one-gene, one-product* definition, this region must be considered to be five genes and a pseudogene.

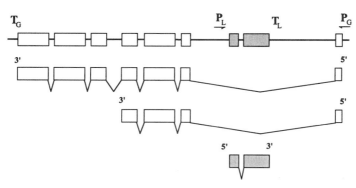

Figure 3. The exon map of the *Gart* locus in *Drosophila*, as described by Henikoff et al. [21]. Although the larval cuticle protein gene *(Lcp)* is fully contained within an intron of the *Gart* locus, its gene is coded on the opposite DNA strand. *Lcp* contains an intron of its own. P_G and T_G are the promoter and terminator for *Gart*, P_L and T_L are those for *Lcp*.

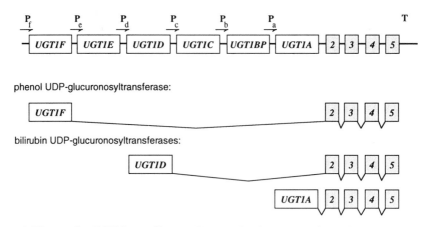

Figure 4. The complex *UGT1* locus of humans [22] contains six promoter sites and one terminator. Each promoter is associated with a separate first exon that is spliced with exons 2-4 to make final mRNAs for translation. The exon labelled *UGT1BP* carries a frame-shift mutation that produces a premature stop codon and is considered a pseudogene.

Complex Post-translational Processing. The human POMC locus produces one large polypeptide from its mRNA. This polypeptide is then processed differentially in different tissue types to give a variety of neuropeptide and hormonal products. In the anterior lobe of the pituitary, the protein is cleaved once, cutting β-lipotropin free from the C-terminal end. The remaining fragment is cleaved again, releasing corticotropin (ACTH) and CLIP (corticotropin like intermediate-lobe peptide). In the intermediate lobe, corticotropin is cleaved again, releasing α-melanocyte stimulating hormone (α-MSH). The β-lipotropin is also cleaved, yielding β-MSH and endorphin. Additional processing yielding additional products also occurs.

How many genes are involved here? Many geneticists seem to prefer the notion that the POMC locus is a single gene. However, under the *one-gene, one-product* definition, this must be considered multiple overlapping loci that happen to share the same transcriptional apparatus.

Although some geneticists might argue that all this is splitting hairs and that the POMC peptide is the product of one and only one gene, others disagree. Alberts et al. [9] note:

> But it is now known that some DNA sequences ... participate in the production of at least two different mRNA molecules and therefore at least two different proteins with distinct biological roles. How then is a gene to be defined? At present, it seem best to retain the one-gene-one-polypeptide-chain definition. This means that in those cases where more than one polypeptide is specified by the same DNA sequence, two or more genes are considered to overlap on the chromosome.

By this definition, the POMC region must be considered at least eight overlapping loci.

Guide RNAs. The most intriguing recent discovery is that some primary transcripts require editing before they become fully functional mRNA molecules. In the mitochondria of trypanosomes, a primary transcript from one transcription unit is modified enzymatically, under the control of guide information contained in another RNA molecule that has been transcribed from a different region of the genome. The resulting mRNA contains information from both original transcripts in a merged form that can then be translated to yield a functional protein. Here the *one-gene, one-product* concept requires that the two separate transcription units be considered one gene.

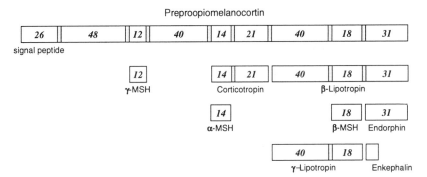

Figure 5. The human POMC locus produces a polypeptide that is processed differentially in different tissues to yield many different functional peptides.

VARIATION IS THE KEY

The study of biology must be the study of variation. Despite the fact that much has been written about sequencing *the* human genome and about obtaining *the* human map, it is generally recognized that considerable genomic variation exists from individual to individual, with estimates that place the amount of sequence difference between individuals at about one nucleotide in three hundred being widely quoted. In addition to base-substitution differences, there are also significant differences in the size of chromosomes, and thus in the actual amount of DNA present. Some have estimated, for example, that human chromosome 1 may show up to 10% differences in length among normal individuals. With that 10% amounting to 30 million bases pairs, the size variance of just one human chromosome may equal ten entire *E. coli* genomes.

With this kind of variation occurring, it is difficult to see how one might really conceive of *the* human sequence and *the* human map, much less represent it as a singularity in a database. Other aspects of normal variation, such as multi-copy genes, also pose challenges for database design.

Multi-copy Genes

By their nature, databases keep track of information relating to individual objects of interest. Databases are not really appropriate for characterizing arbitrary collections of things without individual distinction. Thus, if all genes have equivalent status and if the ultimate genome database is to represent all the genes in the human genome, it will be necessary to identify, name, and describe all the genes. This could be complicated by the occurrence of genes that have clearly recognized function, reasonably straightforward structure, but which occur in the genome as variable-number, multiple copies.

Consider a simple case in *E. coli* where two identical tRNA genes occur as a tandem repeat. It would be easy to give both genes individual names, according to their relative position on the chromosome: $tRNA_1$ and $tRNA_2$. But what if we have to represent the genome of a strain of *E. coli* that has lost one of these genes? Which locus are we to say is missing and which one remains, given that all we know is that there is now only one present and that the two are not otherwise distinguishable, except by their positions relative to each other? Although this may seem a forced example, what if there were hundreds of copies at different locations in the genome? Such is the case with human rRNA loci.

Human cells contain about 200 rRNA gene copies per haploid genome, distributed in clusters on the short arms of the acrocentric chromosomes (chromosomes 13, 14, 15, 21, and 22). Should each of these 200 or so copies be considered an individual gene in its own right? If each copy is a gene in its own right, do all 200 get separate names? The length of the rRNA-bearing chromosomal arms vary significantly among individuals, and thus so presumably do the number of copies of genes. If each gene gets its own name, exactly which of these several hundred named genes does a particular individual carry? And, exactly how many should be placed on *the* human map?

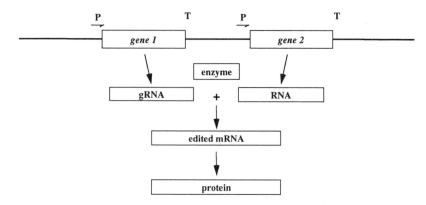

Figure 6. In many mitochondrial systems, it has been demonstrated that RNA editing, under the control of information stored in another RNA molecule (gRNA), is required for the production of a functional mRNA. In this case a single mRNA and a single resulting polypeptide can truly be said to derive from two transcriptional units.

All of these rRNA genes produce equivalent RNA transcripts of about 13,000 nucleotides in length. Each primary transcript is processed in the nucleus to yield one copy each of three different ribosomal rRNAs: 28S rRNA, 18S rRNA, and 5.8S rRNA (Figure 7). According to the *one-gene, one-product* definition of the gene [9], each of these 200 identical transcriptional units must really be considered to be three separate genes. Some believe that the primary rRNA gene transcript yields additional fragments that play a brief functional role during ribosome assembly. If that is the case, then each transcription unit contains more than three genes.

Other Repetitive Elements

Variable Nucleotide Tandem Repeat (VNTR) polymorphisms (Figure 8) have proven to be very useful as mapping reagents. They have also proven to constitute a serious challenge to the notion of a coordinate space on which genes can be placed in the absence of knowledge of what other alleles are present.

It is a given in gene mapping that what one is mapping is the locus at which the alleles of a gene occur, not the particular alleles themselves. It has also been stated many times that "The ultimate, highest resolution map of the human genome is the nucleotide sequence, in which the

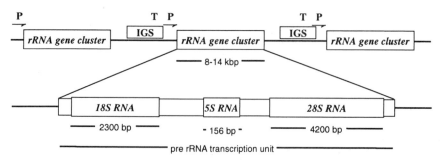

Figure 7. Human rRNA gene clusters occur as variable numbers of tandem repeats on the short arms of the acrocentric chromosomes. Each gene cluster is a transcriptional unit whose RNA transcript is ultimately processed to yield three ribosomal RNAs. "P" indicates a promoter, "T" a terminator site. "IGS" labels an intergenic spacer region. Overall, a haploid human genome carries about 200 copies of the rRNA gene cluster.

identity and location of each of 3 billion nucleotide pairs is known." [19]. This implies that each gene, each functional segment along the DNA, can be identified by the actual address numbers of its first base pair and of its last base pair in the human sequence.

If VNTR alleles can vary in size by thousands of base pairs, and if the most common allele in a population may occur in only a few percent of individuals, then how could we meaningfully assign base-pair numbers as addresses to the loci of VNTR genes, or even to any genes on the other side of a VNTR locus? It cannot be done. There is no coordinate space on which we can pin the location of genes independent of the location of other genes. We must recognize that all genomic mapping must be as offsets relative to other genes known to be present. We must also recognize that errors and uncertainties in the these base-pair offsets will increase as a function of measured distance, so that base-pair-level resolution can only have meaning over relatively short distances.

GENOMIC MAPS OR GENOMIC ANATOMIES?

McKusick [23] has suggested that an anatomic analogy is appropriate when discussing the human gene map: "The anatomic metaphor is appropriate since the linear arrangement of genes in our chromosomes is part of our anatomy. It is also useful for a logical discussion of the significance of the information: the morbid anatomy, the comparative anatomy and evolution, the functional anatomy, the developmental anatomy and the applied anatomy of the human genome."

The anatomy metaphor is also desirable for a more fundamental reason: it is simply better and leads to clearer thinking. The problem with the map analogy is shown in the following story from Richard Feynman's delightful memoir [24]:

> After that I went around to the biology table at dinner time. I had always had some interest in biology, and the guys talked about some very interesting things. Some of them invited me to come to a course they were going to have. ... I had to report on papers along with everyone else [and one of the papers] selected for me ... kept talking about extensors and flexors, the gastrocnemius muscle, and so on. This and that muscle were named, but I hadn't the foggiest idea of where they were located in relation to the nerves or to the cat. So I went to the librarian in the biology section and asked her if she could find me a map of the cat. "A map of the cat, sir?" she asked, horrified. ... From then on there were rumors about some dumb biology graduate student who was looking for a "map of the cat."

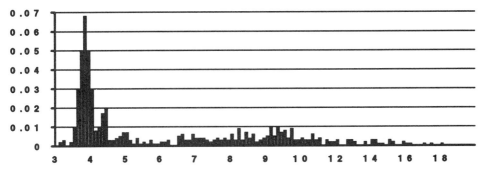

Figure 8. Frequency distribution of *Pst*I-DNA fragments in random individuals from a Caucasian population. Variable nucleotide tandem repeat (VNTR) loci occur in a highly polymorphic form. D14S1 represents the first VNTR studied.

Of course Feynman should have asked for an anatomy of the cat, not a map. But why is his mistake funny? Why is it so obviously crazy to ask for a map of a cat? *The Oxford English Dictionary* defines "map" as:

> **map** *n* A representation of the earth's surface or a part of it, its physical and political features, etc., or of the heavens, delineated on a flat surface of paper or other material, each point in the drawing corresponding to a geographical or celestial position according to a definite scale or projection.

"Mapping," in the mathematical sense, the *OED* defines as:

> **map** *vt* To place (a mathematical aggregate) in a one-to-one correspondence with an aggregate <a set is called denumerable if it can be *mapped* ... onto the set of all the natural numbers --A. H. Wallace>.

We make maps to describe the subcomponents of singular objects: cities, states, continents, galaxies. We make anatomies to describe the average characteristics of sets of objects. It was the failure to recognize that fundamental difference that makes Feynman's request for a cat map so ludicrous.

Genome researchers either laugh or are annoyed when lay persons ask, whose genome will you map? When asked why this is so annoying, they tend to answer, "Because it is so wrong-headed." Some geneticists have been known to respond, "Asking that question is like asking whose face is in *Gray's Anatomy!*" Gray's **anatomy**, indeed.

If someone says that he is making a map of, say, a European country, it is natural to ask which country. The notion of mapping an unspecified singular thing is almost meaningless. Conversely, developing an anatomy based on one specimen is also problematic. If I measure my dog and you measure your dog, how will we ever agree on canine anatomy?

Is insisting that the concept of genomic anatomy is preferable to that of genomic map merely wordplay, or is there something of value involved? I believe that there is indeed something of value. With an anatomy, for example, we may know that some structures are very regular from individual to individual and these are represented very precisely in our anatomical description. Each regular part gets a name and is well described. But some structures may be equally well known to vary considerably from individual to individual. These structures get only generic names, and we take care to point out that much variation is to be expected. No anatomist gives individual names to each of the small venous anastomoses in the human forearm. They are simply too variable to warrant individual names. Similarly, no genome informaticist should be expected to keep track of the many copies of the human rRNA genes, they are simply too variable in number and location. In both cases we must accurately record the variation as perhaps the most important part of the observation.

CONCLUSIONS

What, then, is a gene? Given all of the complex units of regulation and of transcription and of translation that are now know to occur, we must abandon the early molecular concept of the gene as a discrete, contiguous region of DNA with definable function. Instead, we must recognize that a gene or other map object of interest may well consist of a set of not necessarily discrete and not necessarily contiguous regions along a DNA molecule. Only by defining a *set* of regions can we construct data models of sufficient complexity to represent reality. Darnell et al. [10] have said as much:

> The concept of the gene as a biological entity remains intact: a gene is still considered a heritable function detected by observing the effect of the mutation. However, according to the current definition, a gene consists of all the DNA sequences necessary to produce a single

peptide or RNA product. Thus, the gene is no longer thought of a single contiguous stretch of DNA.

These sets must be hierarchical in the sense that permits more complex map-object sets to be created from less complex sets. The set-of-sets concept allows us to represent easily the many regulatory regions of the MMS operon and the participation of the MMS operon in the heat-shock regulon. It also allows us to represent the *UGT1* region as both five separate genes and a pseudogene and as a higher singular order locus consisting of that set of five genes and a pseudogene.

As for mapping, we must abandon the notion of creating a single correct human genomic *map* and begin thinking about developing an accurate genomic *anatomy* instead. This anatomy must be capable of representing regular regions with precision and variable regions with due attention to the variability and uncertainty. Only by recognizing the anatomical nature of the genome will we be able to develop data models and database systems that can carry us through the successful completion of the human genome project.

REFERENCES

1. Cinkosky, M.J., Fickett, J.W., Gilna, P., and Burks, C., 1991, Electronic Data Publishing and GenBank, Science 252:1273.
2. Courteau, J., 1991, Genome databases, Science 254:201.
3. Pearson, M.L., and Söll, D., 1991, The human genome project: a paradigm for information management in the life sciences, The FASEB Journal, 5:35.
4. Carlson, E.A., 1966, "The Gene: A Critical History," W. B. Saunders Company, Philadelphia.
5. Demerec, M., 1933, What is a gene?, J. Hered., 64:369.
6. Demerec, M., 1955, What is a gene -- twenty years later?, Amer. Nat., 89:5.
7. Muller, H.J., 1945., The gene, Proc. Roy. Soc. Biol., 134:1.
8. Stadler, L.J., 1954, The gene, Science, 120:811.
9. Alberts, B., Bray, D., Lewis, J., Raff, M., Roberts, K., and Watson, J.D., 1983, "Molecular Biology of the Cell," Garland Publishing Company, New York.
10. Darnell, J., Lodish, H., and Baltimore, D., 1986, "Molecular Cell Biology," Scientific American Books, New York.
11. Singer, M., and Berg, P., 1992, "Genes & Genomes," University Science Books, Mill Valley, California.
12. Sturtevant, A.H., 1913, The linear arrangement of six sex-linked factors in Drosophila, as shown by their mode of association, J. Exp. Zool. 14:43, reprinted in: "Conceptual Foundations of Genetics," H.O. Corwin and J. B. Jenkins, eds., Houghton Mifflin Company, Boston.
13. Sturtevant, A.H., and Beadle, G.W., 1939, "An Introduction to Genetics," W. B. Saunders Company, Philadelphia.
14. Carlson, E.A., 1959, Comparative genetics of complex loci, Quarterly Review of Biology, 34:33-67.
15. Glass, B., 1955, Pseudoalleles, Science, 122:233.
16. Benzer, S., 1955, Fine structure of a genetic region in bacteriophage, Proc. Nat. Acad. Sci., 41:344.
17. Ayala, F.J., and Kiger, J.A., Jr., 1984, "Modern genetics, Second Edition," Benjamin/Cummings Publishing Company, Inc., Menlo Park, California.
18. Suzuki, D.T., Griffiths, A.J.F., Miller, J.M., and Lewontin, R.C., 1986, "An Introduction to Genetics," W. H. Freeman and Company, San Francisco.
19. Committee on Mapping and Sequencing the Human Genome, Board on Basic Biology, Commission on Life Sciences, National Research Council, 1988, "Mapping and Sequencing the Human Genome," National Academy Press, Washington, D.C.
20. Lupski, J.R., and Godson, G.N., 1989, DNA→DNA, and DNA→RNA→protein: Orchestration by a single complex operon, BioEssays 10:152.
21. Henikoff, S., Keene, M.A., Fechtel, K., and Fristrom, J.W., 1986, Gene within a gene: Nested Drosophila genes encode unrelated proteins on opposite DNA strands, Cell 44:33.

22. Ritter, J.K.,Chen, F., Sheen, Y.Y., Tran, H.M., Kimura, S., Yeatman, M.T., and Owens, I.S., 1992, A novel complex locus UGT1 encodes human bilirubin, phenol, and other UDP-glucuronosyltransferases isozymes with identical carboxyl termini, J. Biol. Chem. 267:3257.
23. McKusick, V.A., 1988, "The Morbid Anatomy of the Human Genome," Howard Hughes Medical Institute, Bethesda, Maryland.
24. Feynman, R.P., and Leighton, R., 1985, "Surely You're Joking, Mr. Feynman!" W. W. Norton and Company, Inc., New York.

LIVERMORE'S PRAGMATIC APPROACH TO INTEGRATED MAPPING FOR CHROMOSOME 19

Tom Slezak*, Mark Wagner, Mimi Yeh, Roy Cantu III, and Elbert Branscomb

Human Genome Center
Biomedical Sciences Division, L-452
Lawrence Livermore National Laboratory
7000 East Avenue
Livermore, CA 94550
U.S.A.

ABSTRACT

For several years Livermore has been using a bottom-up approach to physical mapping, constructing contigs from cosmid clone fingerprints. We currently have 7112 clones placed in 854 contigs and now are using various other means to close gaps, complete, and verify the map. This has required us to develop an information model capable of handling all the important relationships between any "mappable object" and producing a variety of ways of visualizing this integrated data. We have defined a set of relations on a single generic map object class that allows us to distill all the distance/order/orientation/overlap/etc. data contained in our underlying database, along with algorithms to construct partial order trees and display the integrated results in our X-windows database browser tool. This system uses Sybase/SQL/Unix/C and can be used on Sun color Sparcstations with Internet access. We currently have SQL access to the GDB and GenBank database servers but are not yet extracting relevant data from these sources into our integrated map except through our direct collaborations with data submitters. This work was performed under the auspices of the U.S. Department of Energy by Lawrence Livermore National Laboratory under Contract No. W-7405-ENG-48.

INTRODUCTION TO PHYSICAL MAPPING

The Human Genome Project (HGP) is an ambitious multidisciplinary, international effort to locate, characterize, and understand the estimated 100,000 genes that determine the organization and functions of humans. There are an estimated 5,000 diseases of genetic origin that could be affected by this research.

---

*To whom all correspondence should be sent: slezak@llnl.gov, (510) 422-5746, FAX (510) 423-3608

Computational Methods In Genome Research,
Edited by S. Suhai, Plenum Press, New York, 1994

Physical mapping is the use of various biological techniques to isolate the location of genes to specific portions of the 24 chromosomes that comprise the total human genome. The crudest techniques may place a gene only on a specific chromosome, or perhaps on a particular arm or "band" on a chromosome. These regions are still far too large for practical use, since a single band may span several million base-pairs. More sensitive mapping techniques can further refine the location of a gene within a chromosome to areas of clonable size, such as Yeast Artificial Chromosome clones (YACs, from 100,000 to over 1,500,000 base-pairs in length) or cosmid clones (about 40,000 base-pairs long.) A collection of clones specific to a region of interest is called a library. Unfortunately, the process of making a library results in a complete loss of order, thus requiring the use of additional mapping techniques to order the objects on which the genes are eventually mapped.

It should be noted that there are numerous possible different approaches to constructing a physical map, depending on the exact characteristics of the particular type(s) of clones used and the experimental techniques used to order them and map genes upon them. Approaches which rely on the larger YAC clones as their primary objects are often referred to as "top down" approaches, whereas the use of smaller cosmid clones is a "bottom up" method. The HGP is currently employing variations of both strategies on different chromosomes as various biological and computational techniques are being developed and evaluated.

BRIEF OVERVIEW OF THE LLNL MAPPING SYSTEM

Livermore's current role in the HGP [1] is to provide a physical map of human chromosome 19, which at an estimated length of 60 million base-pairs is one of the smaller chromosomes. Surprisingly, it is also considered to be "gene rich", containing an estimated 2,000 genes. As of this writing, probes are available for 90 of those genes and all have been mapped to clones in our library. Included in this set of known genes are ones that cause a form of muscular dystrophy, influence 25% of all heart attacks, and control the sense of smell. Livermore focused on chromosome 19 because of prior work on a set of DNA repair genes that exist on chromosome 19. These genes are capable of locating and correcting certain mutations that are due to the effects of ionizing radiation [2]. It is anticipated that the remaining estimated 1,900 genes on chromosome 19 control many other important functions and may lead to significant advancement in human health knowledge, diagnosis, and treatment.

In the past 5 years we have constructed a system to facilitate the acquisition, storage, analysis, querying, and presentation of all our physical mapping data for chromosome 19. The processing of cosmid clone overlap data comprised the bulk of our work in the early phases. We have now "fingerprinted" over 10,000 cosmid clones, analyzed them for overlap, and reassembled 7,112 of them into 854 contigs that are spanned by 3,057 clones [3]. While we continue to process cosmid clones at a decreasing rate, our efforts now have switched to utilize YAC clones and other techniques to both reduce the number of gaps and to order the cosmid clone contigs. An overview of our entire system (Fig. 1) has recently been written [4].

OUR DNA DATABASE

Since starting our human genome database in late 1989 we have accumulated more than one hundred tables containing over 160MB of data and indicies. In addition, a separate database contains 550MB of nearly-raw cosmid clone fingerprint data (waveforms). We have several SUN workstations with access to the genome databases via our custom browser serving 30 local and several remote end users. Our database has live links to other major genome repository databases (GDB at Johns Hopkins and Genbank at Palo Alto and other satellite sites.) Our database, like these repositories and many other HGP databases, is implemented using Sybase [Sybase, Inc., Emeryville, CA]. We use a Sun Sparc 2 workstation as our primary database server and

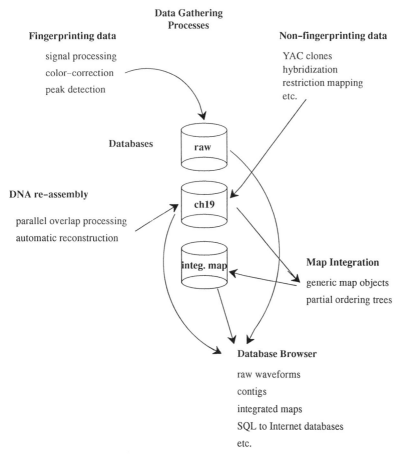

Figure 1. Overall system diagram. This shows the highly-simplified flow of information in our system and the relation to the various database components. The data gathering steps involve over 30 biologists and well over 100 database tables plus the codes to populate and query them. Cosmid clone fingerprint analysis and the subsequent DNA re-assembly has consumed a major portion of our effort over 5 years. Map integration is more recent, simplifying the experimental data complexity into a small set of relations on a single generic map object class. The browser allows us to visualise all the data contained in the three logically separate databases. Over 15 person-years have gone into this system to date.

have several different logical databases to organize our work. The largest, simplest, and least interesting one contains the raw DNA fingerprint waveform data. The main database contains over 100 linked tables of "lab notebook" data and experimental data for chromosome 19. We have a separate small database to control our browser tool, a protected database for collaborators, and several internal databases for development and usage by end-users. Over 200 tables and views are used to support the entire project. All our custom programming uses the C language [5] or AWK [6] on the Unix operating system [Ritchie, et al. (1974)].

Human genome database development at LLNL is done in an iterative fashion that is dictated by the rapidly changing needs of the end users, rather than any formal design for process. The major steps in the development cycle include defining biological requirements data tables, table design, conducting client walkthroughs to confirm the design, table implementation, automating data conversion, designing and implementing customized user interfaces, and user training.

Database table design for the cutting edge of human genome research is much different from designing a database for a well-defined business system. Researchers are constantly creating new objects and techniques or abandoning old ones and sometimes they cannot predict how future experiments will impact existing relations. Due to the constant evolution and flux, we decided to build rapid prototypes instead of trying to build a perfect system at the outset. We have found that our users generally need time to use and understand the data and evolve the best design. Our experience is that 3 iterations are needed, on average, to finalize a particular set of tables. We have evolved techniques which anticipate such frequent change and attempt to minimize its cost.

Our goal for data retrieval is to help our users manipulate their data easily and give them maximum freedom to access stored data in any way that makes sense to them. We note that the lack of an ad-hoc query language in current object-oriented databases makes them unacceptable to us at this time (among other reasons). We have a variety of users with differing levels of database knowledge and computer aptitude. We supply SQL solutions for all common needs and provide training to help users to acquire the necessary skills to be able to query the database themselves via SQL.

It has always been an important part of our philosophy that genome databases should permit easy access for bidirectional data sharing, subject to the constraints of collaborations and current policies on timely publication. In theory, modern relational databases permit control of access permission down to the level of column/row/table/user. In practice, such a fine level of control is nightmarish to set up and administer as collaborations come and go.

We have adopted a model of sharing that currently permits three levels: full collaborators, partial collaborators, and public viewing. Full collaborators would have unrestricted read-only access to all our data, and write access to tables for which they are supplying data. Partial collaborators can only access our data through a controlled version of our browser tool which prohibits unrestricted querying and requires the use of an access-control list to limit viewing of cosmid clone attributes to exactly that subset which they are deemed privileged to see. By giving different access-control lists to different collaborators we can manage many collaborations with a single, simple mechanism. We also have a separate database that contains only data which has been released to the GDB database at Johns Hopkins; a restricted version of our browser is available to allow access to only this data.

It is our opinion that the traditional method of scientific disclosure through peer-reviewed publication needs to be at least supplemented by some form of electronic submission to a public repository database. Such submissions would establish "first publication" of the data, allow for proper referencing of the data in subsequent work, and make publicly-funded genome data available quickly to other researchers. We will be strongly supporting the adoption of the new ISO Remote Database Access (RDA) protocols that will facilitate Internet data sharing over a wide range of hardware and software platforms and urge all Genome researchers to migrate towards this international standard.

Although we utilize all the available techniques to ensure database integrity and security (i.e., separate user accounts, field rules, triggers, read-only views, etc.) we found that it is sometimes necessary to break or bend some of the rules of good database design in order to satisfy practical concerns. We are not dogmatic about normalization. It is more important to us that our end users understand how their data is stored and accessed and that they are comfortable with using the tables. Our users participate heavily in the design of "their" data tables. We also have found it necessary for performance reasons to construct some de-normalized tables to prevent some multi-table joins from causing unacceptably slow response.

MAP INTEGRATION

The first several years of the project at LLNL were concerned almost solely with cosmid clone fingerprinting and re-assembly into spanned contigs. Once we had fingerprinted over 9,000 cosmid clones the emphasis shifted to other biological techniques to both close the 800+ gaps and to order the remaining islands in an iterative fashion until the map was deemed "complete". We will not detail the biological methods used to accomplish these tasks, other than to mention that they involve both large and small DNA objects and a sometimes bewildering array of hybridization schemes to establish overlap, order, and orientation of these objects. Clearly, we needed to develop a mechanism capable of integrating all the diverse data that are being generated, no matter what biological techniques are used.

It is important to note that there are many different sorts of "maps" known to biologists. The LLNL effort is currently only concerned with "physical maps", of which there are numerous different types (e.g., YAC clone maps, cosmid clone contig maps, restriction fragment maps, sequence maps, ordered marker maps, etc.). We are not currently working with "genetic" maps, which involve family studies where phenotypical traits can be observed in those affected with various genetic diseases or disorders. The key observation, from the viewpoint of a systems designer, is that each type of map has a different use and imposes a different viewpoint on the end user. Biologists who deal solely with restriction fragment maps, for example, may have a hard time understanding the issues that a cosmid clone contig mapper finds vital. The systems designer must strive to allow all viewpoints to be equally valid and equally served in the final design. It should also be possible for new viewpoints to be accommodated when new mapping methods are devised.

Data from one type of map may link to data from other map(s) in significant ways (Fig. 2). For example, one important gap in our cosmid clone contig map was spanned by a large YAC clone, which "bridged" two cosmid clone contigs and established their order. This work led to the finding of the muscular dystrophy gene which was announced in early 1992. In short, the sum of the integration of all the available maps will in general be much more than the sum of the individual maps themselves. We needed to design an integration mechanism that could automatically extract maximal order/orientation information over the entire collection of data and be flexible enough to add new objects or relationships as they arise.

Not only are there many different types of maps, there are also many different valid viewpoints for thinking about integrated maps. Almost every biologist has a slightly different idea about what the "ideal" integrated map would look like and how much detail must be shown. Once again, we decided that a useful end goal would be to try to build a system that was flexible enough to be capable of meeting any rational demand for dealing with integrated data. This implied that the display of the maps must be separate from the integration of the data.

It is also important to recognize that there are multiple different categories of potential users of integrated maps, each with their own concerns and needs. Labs that are producing maps, like LLNL, need access to a tremendous amount of detail data in order to adequately control the design of future experiments. Producer labs also must deal with conflicting data, rapid changes, and the higher error rates that accompany oven-fresh data. Central repository databases, of which GDB is a good example, simply cannot tolerate great amounts of detail, flux, and uncertainty. Similarly, the needs of an on-line map viewer will be vastly different from a user who wants to produce some form of a hard-copy map. The frequency of map updates may also be an issue.

It is clear to us that the existence of multiple valid uses and viewpoints precludes a single, "best" solution for integrated mapping that will meet all needs of all users at all times. We chose an approach that fit the needs of our users while providing flexibility for future changes in both data and display needs.

In a somewhat radical departure from conventional wisdom, we decided that there was no compelling reason to store an integrated map in the database as a geometric model of the final physical representation. In other words, to us a genomic map is a complex web of relationships between objects, with infinite possible views of subsets of those relationships. This argues strongly against using any data model for integrated maps that makes any assumptions about the method(s) to be used to display the maps, and strongly for using a model that is capable of supporting any possible combination of views and display methods. Such an approach simplifies the process of map integration and storage and provide a clean separation between the integration and display processes.

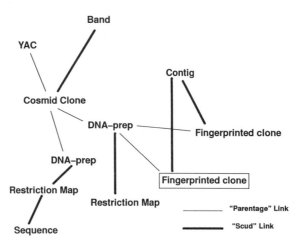

Figure 2. Integrated Data has complex linkage. Complex linkages. This diagram hints at the web of relationships that can exist between mappable objects. A YAC clone might be sub-cloned into a cosmid clone, which could be insitu hybridised to a band and might then have multiple DNA preps made. Each of these might have one or more restriction digests performed, possibly with sequencing being done on the fragments, and one or more cosmid fingerprints analysed. The analysed fingerprints are reassembled into (hopefully a single) contig. Note that a "parentage" linkage keeps track of which objects are derived from (essentially) the same source. All other linkages, primarily hybridisations or other membership relationships, are maintained by the "scud" relation. We desire to be able to instantly see fall the information shown in this diagram if we were to focus attention on any member of the parentage tree.

A GENERIC APPROACH TO MODELING MAP INTEGRATION

It is precisely the issue of the amount of detail data that should be shown that leads to the myriad viewpoints of integrated maps ("I want to see these sorts of detail on these N regions I am concerned with, but only summary data everywhere else.") We decided to approach the problem of map integration by observing that, at one level of thinking, all the "parts" of an integrated map were equivalent to every other part. That is, each generic map object has a physical length and/or a physical location on the completed chromosome (although at present either or both of these might be unknown or known only grossly). Furthermore, some objects might be known to be derived from some other "parent" object, or might be known to overlap one or more other objects (i.e., from hybridization tests.) The order of some objects might be known (the spanning set of a contig for example) and the orientation of 2 objects relative to each other could be established (clone A is located on the P arm telomeric to clone B.)

We note that our model for map integration is greatly influenced by both the mapping method(s) that we depend on most heavily and the type of data that our main experimental method(s) provide (i.e., cosmid clone contigs with minimal spanning paths defined.) The model must meet all the local needs and account for any experimental peculiarities. However, this generic approach to modeling integrated maps makes it simple to customize for such peculiarities.

In short, only a few generalized relations that act upon a single class of generic map objects are sufficient to allow us to coalesce all the richness of data that is contained in over 100 tables of our underlying experimental database. As shown in Figures 3-5, we have defined a set of 10 such generalized relations to describe our integrated data. Most of these are rather obvious (absolute location, location relative to some other object, length, distance between two objects, overlap between two objects, orientation of two objects) but some require a bit more explanation. We have a parent-of relation to accommodate the fact that our DNA preps come from cosmid clones and our fingerprints come from DNA preps, etc., which allows us to readily locate all objects which are the "same" in some abstract sense. Similarly, we have an all-purpose relation nick-named "scud" that allows us to group any sort of experimental information that claims that two objects share some common region(s) of DNA. Hybridization results are the most obvious example of this relation. We have a "tri-order" relation that was used to store fluorescent insitu hybridization (FISH) data that stated that object A either was or was not between objects B and C. As we are now using distance-based techniques for ordering, this relation may be dropped in the future. Finally, we have a concept of meta-objects embodied in our "span" relation, which allows us to define an ordered (but possibly unoriented) set of overlapping objects which could be of any underlying type (including, perhaps, other spans). This relation is now being exploited as we use YACs and other objects to link and order our cosmid contigs. We have made no attempt to define the absolute minimum number of relations needed to encompass all information needed to do map integration. For example, our automatic reassembly algorithm gives us spanned contigs and we choose to store those spanning paths in the map object database rather than infer them a second time from the overlap information. We will leave the question of the minimal necessary set of relations to others and note that 9 or 10 basic relations are sufficient to get the job done, given our current experimental data.This approach to integrated mapping allows us to run fully-automated partial ordering algorithms across all data to assemble all order and orientation information, even though the data may originally come in bits and pieces from individual relations in many different experimental database tables.

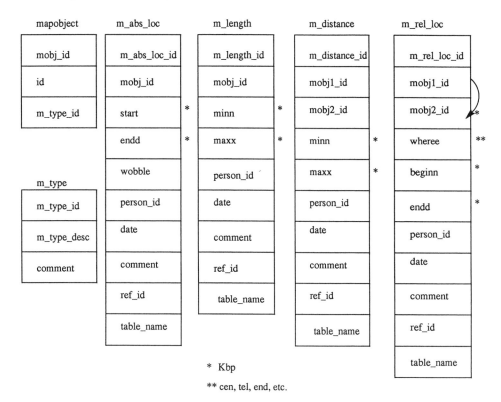

Figure 3. Schematic diagram of map object relations. Note that all relations in Figures 3-5 deal solely with generic map objects. The odd spellings of some field names are due to SQL keyword clashes. The m_rel_loc relation places mapobject 1 somewhere on mapobject 2, whereas the m_abs_loc relation places a map object at an absolute location relative to a (mythical) P-terminal zero origin.

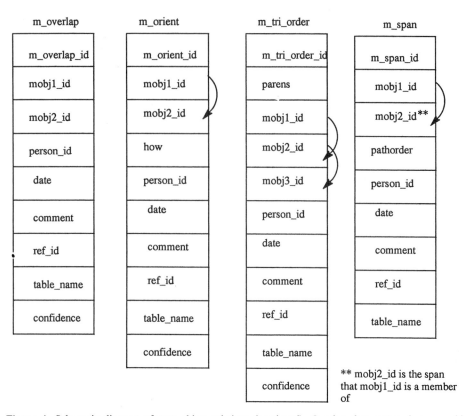

Figure 4. Schematic diagram of map object relations (continued). Overlap data comes from cosmid fingerprint analysis, but this table could handle any other sorts of pair-wise overlaps that might be developed. The order and orient relations feed the partial ordering algorithm with pair-wise facts, using an orientation convention that object 1 is P-terminal of object 2. The tri-order relationship handles FISH data that placed one object either between or not between two other objects. As we decrease the use of this type of FISH data, this table is becoming obsolete. The span table allows us to define meta-objects (i.e., meta-contigs) that include overlapping objects of various underlying types. This is vital for using YACs to span cosmid contigs, for example.

We note that our approach defies the current conventional wisdom, which seems to say that it is "obvious" that this problem demands an object-oriented solution. We feel that such an approach puts too much emphasis on the attributes of the various mappable objects and too little emphasis on the relationships between objects (or, makes it difficult to extract relationships across various "classes" of mappable objects.) We feel that our "relation-oriented" approach offer more direct access to the data needed for integrated mapping. In our system it is a trivial query of a single relation to answer the question "Give me all objects that are known (via any method) to lie on YAC 1234", regardless of the type of objects (cosmid clone, gene, locus, genetic marker, restriction digest fragment, etc.) We feel it is highly probable that the hierarchical structure imposed by current object-oriented systems will discourage such sweeping queries or require unnatural contortions to accommodate them.

IMPLEMENTING THE GENERIC MAP OBJECT DATABASE

We implemented a set of fully-automated script procedures and C programs to transform all relevant data from the experimental database into the generic map object integration database tables (Fig. 6). The "mapper" portion of the browser, as described below, reads and

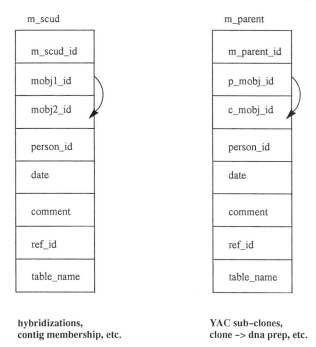

m_scud

m_scud_id
mobj1_id
mobj2_id
person_id
date
comment
ref_id
table_name

m_parent

m_parent_id
p_mobj_id
c_mobj_id
person_id
date
comment
ref_id
table_name

hybridizations,
contig membership, etc.

YAC sub-clones,
clone -> dna prep, etc.

Figure 5. Schematic diagram of map object relations (continued). The scud (Securely Confirmed Ultimate Destination) relation was coined during the Gulf War, due to the perceived similarity between a missile of that name and the many hybridization techniques that are used to blindly probe the YAC and cosmid libraries. It also allow us to express concepts like contig and restriction map membership, giving us a generic way to express that object 1 has at least some DNA purported to be in common with object 2. The parent table expresses a similar relationship, but the pedigree is known for certain since this is an "is derived from" relationship. Whether this distinction is necessary is a matter of opinion.

manipulates these tables to present a variety of integrated map display formats. We currently have about 114,000 generic map objects in the database, with a total of 152,000 relations defined between those objects. It takes one hour to populate the map object tables from the underlying experimental data.

Our partial ordering algorithm only uses the map integration tables that contain order and orientation data. This algorithm puts that data into the matricies and iterates until either they are empty or until it determines that no more trees can be generated and that one or more unused facts remain. It constructs adjacency and maximal path length matricies and uses these to generate compacted partial order trees that are either oriented or unoriented, depending on whether orientation information is known about any 2 tree members. The algorithm also detects and prints cycles so that errors or inconsistencies in the underlying data can be examined and corrected (perhaps by running more experiments to resolve conflicts). The power of the generic map object approach is demonstrated by the fact that a single, simple partial ordering algorithm can operate over the sum of order/orientation information known for the entire map and extract the best possible information, regardless of the original experimental source of the data. This process runs in about a minute on the data mentioned above. This algorithm has been used to help determine the most optimal experiments to perform to resolve the ordering of a large set of markers on chromosome 19. We are modifying our process to extract ordering information from new data that only gives us the distance between pairs of objects.

An unexpected benefit of this generic approach was our realization that the method which works for the production of maps-in-progress could also be of value in producing consensus maps from a variety of independent sources. All that would be needed is the use of the same name

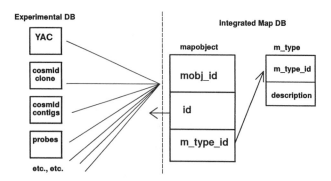

Figure 6. "Parts is parts": Generic Map Objects. Each mappable object in our experimental database becomes an instance of the generic map object class. The database key from its underlying experimental object is retained, along with a reference to the underlying map object type. This compression of irrelevant detail allows us to compress all salient mapping information into less than a dozen tables.

(official probe names, etc.) for identical objects across all input maps. A mapping table could then map all objects to generic map objects, and all salient relationships from the input maps could then be translated into the basic set of relationships on these pooled generic objects. The partial ordering algorithm would then be run on this merged set of relations and a consensus of ordering would emerge (and likely some conflicting data as well.)

BROWSING THE MAP

The implementation of our database browser (a graphical tool for viewing and navigating our database) has been an ongoing effort for over three years, with researchers using it on a daily basis. Our browser was designed to be a flexible framework, capable of adding new data types and display methods as they were requested. Each different way of looking at physical maps is valid, and should be at least potentially allowable in our system. Each different view of our data can be thought of as an independent "display object."

We designed the browser to have a flexible framework consisting of two displays, one above the other. Each of these "display object stacks" is completely independent of the other, so the user has the ability to look at one view of the chromosome in one window while looking at another view in the other window. When the user selects something to look at, a new "display object stack frame" is placed on that display stack. When the user is done with that window, they can pop the top frame off the stack, and be back where they were previously. This permits the user to see something of interest, investigate it fully, then return to where they were previously with a minimum of effort. There are many other valid possible design choices (infinite pop-up independent windows, for one.)

There are times when the half-height display stack windows are not large enough to display all the available information. To more effectively utilize the finite amount of screen space, we have provided the user with the capability to "double" any display window. This expands the selected choice so it fills the space of both display windows, giving the user twice as much room to view things as before. It also allows us to better use the screen for large display items where the half-height screen would not be adequate.

One of the most important features of the browser is what we call the query menu (Fig. 7). We needed to satisfy a range of users, from total computer novices to experts. We had to design an interface that would neither be intimidating nor frustrating to both groups. We also wanted to not limit the types of queries that could be made. We accomplished this with a generic user interface. The user can select from a list of "canned" queries which do the most common tasks. The user can also type in direct SQL through an editing feature. They can use a canned query as a template, and then edit that query to suit their current needs. Canned queries are stored in the database, as are all other data that customize the query menu for each browser display method. The query menu code knows nothing about biology or SQL. It only knows several methods of allowing the user to construct queries, which are then passed blindly to the SQL server. Return values are assumed to be database keys of a type that is valid for the current display mode (e.g., mapobject_id for the Integrated Mapper).

Besides canned queries that take user-supplied values at run time, our interface also allows user queries to be read from and stored to arbitrary files. Not all desired functions can be done with SQL alone (e.g., recursive queries) so we had to allow for arbitrary external processes (or pipelines of processes) to be run and their output to be a stream of database keys that could feed directly as a "hit list" to the query menu interface. The user may have generated a list of items of interest and put these items into a file, or may have a shell script or program that generates a list of items. The query menu can read such lists directly and provides a reasonable default display when presented with a null list.

We have been requested to implement 5 independent ways to look at the database (so far) but will only discuss the Integrated Mapper here.

The Integrated Mapper allows us to step back and see more global relationships between the biological entities (Fig. 8). The Mapper looks at the relationships between entities at different levels (i.e., which Yeast Artificial Chromosomes was this clone sub-cloned from, or to which band has this clone been linked.) Consider the many different types of items in our data hierarchy and the vast differences in the scale of those objects. If we were to represent a base-pair of a sequence as a single pixel on a 100 dots/inch display we would need over 9 miles in length to represent the 60,000,000 base-pairs of chromosome 19! It should be clear that panning and zooming alone are insufficient to handle the scale problem. Our users are primarily interested in using the Mapper as a tool to view work in progress, so we emphasize the display of data relationships, not graphic artistry, in our present display modes.

The Mapper also shows how we continued to use our basic design philosophy when faced with constant change. We could not make any assumptions about how users would prefer to view relationships, so we supplied a general mechanism. At present there are 14 different types of mappable objects displayable. New types can be added easily. The user can always see how many of these types have items to show based on the current point of focus and can specify which 1-4 objects will be viewable at any instant (a screen real estate limitation.)

The user may want to navigate in this sea of relationships, so we have created a stack of subwindows within the Mapper itself. The Map stack permits flexibility: Duplication of the top item on the stack allows us to see something else and get back to this point. A simple exchange mechanism permits the user to quickly toggle between two items or areas of interest.

The browser is constantly evolving. More objects will be added to the data hierarchy and new display methods will be developed as our work on chromosome 19 progresses.

SUMMARY

We have developed a simple data model for integrated physical maps of DNA that utilizes a single class of generic mappable objects and a small set of fundamental relations between such objects. We populate this database automatically from the underlying experimental database at regular intervals. We use a global partial ordering algorithm to extract maximal order/orientation information from the aggregate map object database. Our data model makes no assumptions about how the maps will eventually be displayed. That task is handled by a portion of our database browser tool, which employs several mechanisms to allow the user to customize the presentation to meet specific needs. We feel that this approach might prove useful to others who are constructing integrated physical maps.

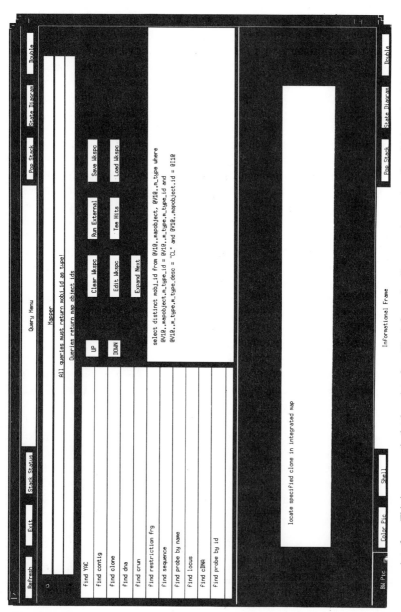

Figure 7. Browser query menu interface. This is our generic database interface. The information lines near the top describe which mode of the database browser we are in and what type of database keys must be returned by all queries. A scrollable list of canned queries is on the left, with a mouse-click the user can get a description of the query or pull up the SQL template in the workspace on the right. The query to find a specific clone in the integrated map has been selected. Note the embedded markers in the SQL template indicating that the user must interactively supply various values, by clicking the "Expand Next" button. Once the query is fully expanded, the red "Run Query" button is hit (invisible in this B/W picture, it is to the right of the "Expand Next" button.) Other buttons allow an editor window to appear, with the workspace contents pre-loaded, and for external queries to be run using either the workspace contents or specified external files. Queries can be saved to files or loaded from files.

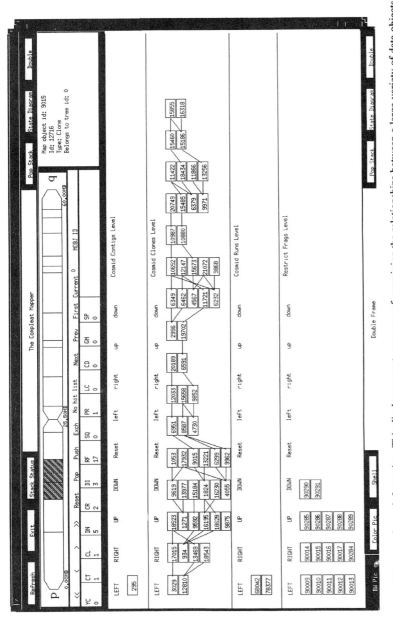

Figure 8. Browser view of integrated map information. This display presents one way of examining the relationships between a large variety of data objects. Near the top, a diagram shows the chromosome and the band(s) associated with the current object of focus. Below the diagram are controls for scrolling and manipulating the map object stack, as well as controls for the query hit list (if any). Below that are the indicators to control the display of up to 4 of the available map object types at any time (13 types are shown in this image, which does not show the use of color to indicate object selection.) We have chosen to display contigs, clones, "Cosmid runs" (fingerprinted and analyzed clones), and restriction fragments. An information window at top right shows that object 9015, which is clone id 12716, has been selected for focus (the red color used is lost in this rendering.) This clone is in a partial-order tree in the cosmid clone display, and has links to a contig, 2 cosmid-run objects, and 17 restriction map fragments. We can see from the object selection boxes that there is data for this clone in 3 other levels that we could choose to show, by turning off one or more of the current views.

109

ACKNOWLEDGMENT

This work was performed under the auspices of the U.S. Department of Energy by Lawrence Livermore National Laboratory under Contract No. W-7405-ENG-48.

REFERENCES

1. Carrano, A.V., et al. (1989). A high-resolution fluorescence-based semi-automated method for DNA fingerprinting. Genomics 4, 129-136.
2. Weber, C.A., Salazar, E.P., et al. (1988). Molecular cloning and biological characterization of a human gene, ERCC2, that corrects the nucleotide excision repair defect in CHO UV5 cells. Mol. Cell. Biol. 8, 1137-1146.
3. Branscomb, E., et al. (1990) Optimizing Restriction Fragment Fingerprinting Methods for Ordering Large Genomic Libraries. Genomics 8, 351-366.
4. Slezak, T., et al. (1992) A system for constructing, integrating, and displaying a physical map of chromosome 19. Submitted to 26th Hawaii International Conference of System Sciences, Jan 5-8, 1993.
5. Kerninghan, B., and Ritchie, D., (1988). The C programming language, 2nd Edition. Prentice Hall.
6. Aho, A., et al. (1988) The AWK Programming Language, Addison-Wesley.
7. Ritchie, D., and Thompson, K. (1974) The UNIX timesharing system. CACM 17, 365-375.

SEARCHING PROTEIN SEQUENCE DATABASES - IS OPTIMAL BEST?

William R. Pearson

Department of Biochemistry
University of Virginia
Charlottesville, VA 22908

INTRODUCTION

The explosion of DNA and protein sequence data over the past 10 years and the advent of the Human Genome Project has produced tremendous interest in the best methods for identifying distantly related genes. Methods for calculating mathematically rigorous alignments of biological sequences have been known for more than twenty years. Because they are optimal in a mathematical sense, the rigorous methods are often considered best for biological problems. Until recently however, these methods have been too slow for comprehensive protein sequence library searches and they remain impractical for exhaustive DNA library searches. The molecular biological community has increasingly relied on more rapid, heuristic approaches like the FASTA and BLAST algorithms. One is concerned, however, that these rapid approximate techniques may miss biologically significant relationships.

In one biological context, the best sequence comparison algorithm is the one that allows us to find those protein sequences that have diverged the farthest from their common ancestor. Such a criterion is appropriate for identifying a newly sequenced protein by comparing it against every sequence in a protein database. For this problem, a very sensitive algorithm might calculate a high similarity score for human hemoglobin alpha and a plant leghemoglobin, while a less sensitive algorithm might calculate the same score for hemoglobin/leghemoglobin as for two unrelated sequences, e.g. hemoglobin/trypsin. Table 1 summarizes a variety of sequence comparison algorithms that have been used for evaluating protein sequence homologies. The first algorithms to be described, those of Needleman Wunsch[1] and Sellers,[2] calculate a "global" similarity score that is better suited for evolutionary tree construction than for similarity searching (Table 2). In this paper, we will focus on algorithms that calculate a "local" similarity score; the most widely used "local" algorithms are the Smith-Waterman algorithm,[3] the FASTA algorithm,[4] and the BLAST algorithm.[5] Table 2 shows the scores calculated by different sequence comparison algorithms for the rat M1 muscarinic acetylcholine receptor when it is compared to two other members of the G-protein- coupled receptor family and to an unrelated cytochrome P450 sequence.

Our discussion focuses on protein sequence comparison, because protein sequence similarity can reliably be used to identify sequences that diverged more than 2.5 billion years ago.[8] In contrast, DNA sequence comparison of nonprotein-coding sequences can rarely be used to look back more than 100 million years.

Computational Methods In Genome Research,
Edited by S. Suhai, Plenum Press, New York, 1994

Table 1. Sequence comparison algorithms

algorithm	value calculated	scoring matrix	gap penalty	time required [1]	ref.
Needleman-Wunsch	global similarity	arbitrary	penalty/gap $q + rk$	$O(n^2)$	1
Sellers	(global) distance	unity	penalty/ residue	$O(n^2)$	2
Smith-Waterman	local similarity	arbitrary	penalty/gap $q + rk$	$O(n^2)$	3 6
FASTP FASTA	approx. local similarity	arbitrary	limited gap size $q + rk$	$O(n^2)/K$	7 4
BLAST	maximum segment score	arbitrary	no gaps	$O(n)$	5

[1] Modern versions run in this time.

In the context of identifying distantly related proteins there is also considerable interest in the best search parameters; i.e. the best matrix of scores for amino acid identities and replacements and the most appropriate values for the start of a insertion or deletion and additional residues in the insertion/deletion (gap penalties). The most widely used substitution matrix is the PAM250 matrix [9] which is derived from amino acid substitution frequencies determined from the modest amount of sequence data available in 1976. New substitution matrices have been proposed by Henikoff and Henikoff [10] Taylor et al [11] and Gonnet et al.[12] that are based on a much larger set of protein sequences. Gap penalties are especially problematic. Modest changes in gap penalties appear to cause substantial changes in the implied structural alignment and, while replacement scores have a foundation in the probability of change from one residue to another, there is no analogous theoretical basis for gap penalties.

METHODS

To identify distantly related protein sequences, one must balance sensitivity - the ability to recognize, or calculate high similarity scores for distantly related sequences - and selectivity - the ability not to calculate high scores for unrelated sequences [13]. An algorithm like BLAST,[5] which uses statistical estimates to focus on the most significant alignments between two proteins and does not allow gaps, may perform better for than the more rigorous and sensitive Smith-Waterman algorithm[14] with some protein families, because, although it may miss the most distantly related sequences, it very rarely calculates high scores for unrelated sequences. Thus, to characterize the performance of different sequence comparison algorithms or scoring matrices, one must use a measure that relates the scores of the lowest-scoring related sequences to the highest-scoring unrelated sequences.

To characterize the best search algorithms and the best scoring parameters, we have compared the performance of an algorithm or scoring matrix when it is used to identify all the members of a protein superfamily using a single member of the family as a query sequence [15]. We have used the superfamily classifications provided by the annotated section of the PIR International protein sequence database (PIR1, release 32, March, 1992) [16]. Members of a superfamily are believe to have diverged from a common ancestor; the inference of superfamily membership may be based on structural and functional similarities as well as sequence similarity. We use the superfamily annotations to classify each sequence of the 9633

Table 2. Similarity scores calculated with different algorithms. Each sequence was compared to the rat M1 muscarinic acetylcholine receptor (PIR code A29514) using the algorithm indicated. FASTA comparisons were performed with the ktup parameter set to 2 or 1, and both the unoptimized (initn) and optimized score is shown.

algorithm	clearly related dopamine D2[1] receptor	distantly related thromboxane A2[2] receptor	unrelated cytochrome P450 4A8[3]
Needleman-Wunsch	402	-9	-100
Smith-Waterman	469	120	73
FASTA ktup=2	378	57	34
ktup=2/opt	469	120	50
ktup=1	424	304	113
ktup=1/opt	469	120	73
BLAST	163	63	33
(p-score[4])	744	108	0

[1]PIR entry S08146. [2]PIR entry S13647. [3]PIR entry A36304. [4]-ln(P).
Values for both related sequences are highly significant.

(3 143 580 amino acid residues) in PIR1 to or unrelated to the query sequence. We then searched the PIR1 library with a member of each of the 69 superfamilies with twenty or more members. A list of the query sequence names, the corresponding superfàmily, and the number of superfamily members in the PIR1 database (size) is shown in Table 3.

The results shown below were obtain using an Intel iPSC/860 Hypercube parallel computer running the PCOMPFA and PCOMPSW programs.[17] Typically, 32 i860 nodes were used. Runs with FASTA, $ktup = 2$ required about 1 minute of elapsed time; FASTA, $ktup=1$ required 4.5 m; FASTA, $ktup=2$, optimizing all sequences in the library, 4 m; FASTA, $ktup=1$ with optimization, 7.5 m; Smith-Waterman required about 90 m. The BLAST program was run on a conventional high-performance workstation.

In our previous work, the measure that we used to characterize sensitivity and selectivity was the number of related-sequences with similarity scores less than the score of the 99.5% of the unrelated-sequences.[15] In effect, we considered a related sequence missed if it could not be found after examining the 48 (0.5% of 9633) highest scoring unrelated sequences. We have developed a slightly different criterion for measuring performance that balances the number of related-sequences missed and unrelated-sequences found. The results in this paper are based on the number of related superfamily members that obtained scores lower than the equivalence score. The equivalence score is the score that balances the number of unrelated sequences obtaining scores higher than the score and the number of related sequences obtaining scores lower than the equivalence score. Thus, in a search the globin superfamily with human a-globin using the Smith-Waterman algorithm (Table 4), there were 17 members of the globin superfamily that obtained scores less than or equal to 73 and 17 non-globin sequences that obtained scores that were greater than or equal to 73.Unlike the 0.5% criterion that we used in previously,[15] the equivalence score reflects the distributions of both the high-scoring unrelated-sequences and low-scoring related-sequences.

Table 3. Superfamilies examined

Query	Description/Superfamily	Size
GGLMS	globin V	456
K1HUAG	lamprey human Ig k-chain V-region	268
CCHU	human cytochrome c	140
N2KF1U	alpha-bungarotoxin/snake toxins	109
OKBOG	cGMP dep. protein kinase	107
FEPE	bacterial ferredoxin	91
TPHUCS	human troponin C	86
K3HU	human k-chain C-region	74
RKMDS	ribulose-bisphosphate carboxylase (small chain)	73
HMIVV	influenza A hemaglutinin	72
HLHUB2	class-I HLA alpha chain	69
CYBOA	bovine alpha-crystallin	64
IPHU	human insulin	64
PSHU	human phospholipase A2	57
GNFV1R	RSV pol polyprotein	45
TVHURA	human n-Ras transforming protein	43
DEHUGL	glyceraldehyde-3-phosphate dehydrogenase	41
TRRT1	rat trypsin	40
NRBO	bovine pancreatic ribonuclease	40
AJHUQ	human glutamate--ammonia ligase	39
IVHUI6	human interferon alpha-I-6	39
VCLJB	env polyprotein / type C retrovirus	35
AZBR	azurin / plastocyanin	35
PWHUA	H[+]-transporting ATP synthase (alpha chain)	35
O4HUD1	human cytochrome p450 2D6	33
ATHU	Human skeletal actin	33
KRHUE	human 50K type I keratin	32
HSHUA1	human histone H2A.1	31
GCBO	bovine glucagon	30
VPXRWA	human rotavirus outer capsid protein vp8	29
VHIV34	influenza virus nucleoprotein	27
LZHU	human lysozyme	27
HSHUB1	histone H2B.1	27
NTSRIA	scorpion neurotoxin I"	26
FOVWH3	HIV I gag polyprotein	25
UBPGA	pig tubulin alpha chain	25
XHHU3	human antithrombin-III	25
HNNZS	sendai virus hemaglutinin-neuraminidase	25
VGXR1S	simian rotavirus glycoprotein vp7	25
TVHUM	human myc transforming protein	24
CFKKA	C-phycocyanin alpha chain	23
W6WL18	human papillomavirus E6 protein	23
JDVLVH	hepatitis B DNA-polymerase	23
MNIV2K	influenza A nonstructural protein NS2	22
ACHUA1	human nicotinic acetylcholine receptor	22
KIBET	herpes thymidine kinase 1	22
MNIV1K	influenza A nonstructural protein NS1	22
GNWVY	yellow fever virus genome polyprotein	22

HSHU1B	human histone H1b	22
TIBO	bovine basic proteinase inhibitor	22
KIHUG	human phosphoglycerate kinase	21
UART	rat alpha-2u-globulin	21
SMHU2	human metallothionein	21
W1WLE	human papillomavirus E1 protein	21
W2WLE	human papillomavirus E2 protein	21
SAVLAH	hepatitus B major surface antigen	21
P2WL	human pappillomavirus L2 protein	21
TYTUY2	tuna protamine Y2	21
P1WL	human pappillomavirus L1 protein	21
GNNY1P	human poliovirus genome polyprotein	20
VGNZSV	sendai virus fusion protein	20
HSHU33	human histone H3.3	20
LNHU1	human hepatic lectin H1	20
W7WLHS	human papillomavirus E7 protein	20
VCLJH3	HIV 1 env polyprotein	20
QRECBD	E. coli B12 transport protein btuD	20
PWHU6	human mitochondrial H[+]-transporting ATP synthase	20
DEHUAA	human alcohol dehydrogenase	20
PPPA	papaya papain	20

RESULTS

Comparison of the BLAST, FASTA, and Smith-Waterman algorithms

A comparison of the performance of BLAST, FASTA, and the Smith-Waterman algorithms is shown in Table 4.   Only 36 of the 69 query sequences are shown; the results with the other 33 sequences for each of the algorithms differed by only one or zero. However, it is reassuring to learn that the FASTA program in its fastest mode (ktup=2) performs as well as the Smith-Waterman algorithm, which is 100-times slower, for 41 of the 69 superfamilies tested and, with *ktup=1*, performs as well on 50 of the superfamilies. We will show below that if FASTA optimizes all of the similarity scores, it performs as well as Smith-Waterman. While the results in Table 4 allow us to begin to evaluate the different comparison algorithms, for some cases, it is difficult to know whether the differences in performance are significant. In particular, it is unclear whether missing 156 globin-family members with FASTA, *ktup=2* is substantially different from missing 29 with BLAST, it may be that there is a cluster of related globin sequences which, if FASTA misses one of these sequences, it will miss all of them. This problem could be minimized by pre-processing the library to remove all sequences that shared more than 80%, or 50%, or perhaps 30% identity, but it is not clear whether by doing so, one could guarantee that the finding of a library sequence with a query sequence would be independent of the finding of another superfamily member.

We can assume confidently that the performance on each superfamily is independent of the performance on each of the other superfamilies. Thus, one can tabulate the data by recording only whether one of two comparison algorithms (or scoring matrices) missed more sequences, fewer sequences, or the same number. Once the data are presented in this form, we can use the sign test to determine whether each of the two strategies performs equally well; i.e., the probability p that strategy 1 performs better than strategy 2 equals $q=1-p$, the probability that strategy 2 will perform better than strategy 1. We can calculate from the binomial distribution the probability that the observed numbers of +'s and -'s will occur under the null hypothesis that $p = q = 0.5$. For this data, we approximate the binomial distribution by the Gaussian distribution for the large numbers of superfamilies

Table 4. Library Search Sensitivity

Query Sequence	Superfamily	Family Size	Number of related sequences missed (equivalence)			
			BLAST	FASTA ktup=2	ktup=1	Smith-Waterman
Smith-Waterman better than ktup=1						
GGLMS	globin	456	29	156	26	12
K1HUAG	Ig V-k-chain	268	22	90	59	16
OKBOG	cGMP dep. kinase	107	17	20	20	3
TPHUCS	E-F-hand	86	6	16	12	6
K3HU	Ig C-k-chain	74	17	22	22	16
HMIVV	hemagglutinin	72	9	11	7	3
HLHUB2	class I HLA	69	7	15	2	0
CYBOA	alpha-crystallin	64	5	5	5	3
TRRT1	serine protease	40	15	14	17	10
AJHUQ	glut.-NH3 ligase	39	6	9	8	2
VCLJB	env polyprotein	35	8	11	11	4
AZBR	plastocyanin	35	25	24	26	21
O4HUD1	cytochrome p450	33	1	6	3	1
KRHUE	type I keratin	32	1	4	4	2
GCBO	glucagon	30	5	11	12	12
NTSRIA	neurotoxin I"	26	5	8	7	5
UART	a-2u-globulin	21	13	13	8	6
LNHU1	hepatic lectin H1	20	7	9	9	6
QRECBD	B12 transport btuD	20	6	11	6	4
PWHU6	H+-trans ATP synth	20	9	8	2	7
ktup=1 equivalent to Smith-Waterman						
CCHU	cytochrome c	140	19	28	26	25
N2KF1U	snake toxins	109	9	5	5	4
GNFV1R	pol polyprotein	45	7	8	6	5
XHHU3	antithrombin-III	25	2	4	2	1
W6WL18	papilloma E6	23	2	6	0	0
TIBO	trypsin inhibitor	22	1	3	1	1
SMHU2	metallothionein II	21	3	6	4	5
W7WLHS	papilloma E7	20	1	4	0	0
DEHUAA	alcohol DH	20	0	2	0	0
ktup=2 equivalent to Smith-Waterman						
FEPE	ferredoxin	91	52	53	53	53
IPHU	insulin	64	3	3	3	3
PSHU	phospholipase A2	57	2	2	2	2
HSHUA1	histone H2A.131	0	1	2	1	
FOVWH3	HIV I gag polyprotein	25	4	2	2	2
MNIV2K	influenza virus NS2	22	2	2	2	2
HSHU1B	histone H1b	22	3	2	2	1
TYTUY2	protamine Y2	21	2	2	2	2

(> 10) that differ.

Viewed as +'s and -'s, the results obtained with BLAST and FASTA (*ktup=2*) in Table 4 can be viewed as 4 +'s and 27 -'s; i.e. FASTA performed better than BLAST 4 times and worse 27 times. When differences of ±1 are included also, the values become 10+ and 27 -, which corresponds to a Z- value of 2.79 and has a P < 0.005 in a two-tailed test (i.e. the test is that the two strategies perform differently, not that one strategy performs better than the other). Thus, BLAST performs significantly better than FASTA, ktup=2.

Table 5 compares the Smith-Waterman algorithm (slowest) and the BLAST algorithm (fastest) with FASTA using several parameter settings. The (120) or (250) indicates whether the PAM120 amino acid substitution matrix (the standard matrix for BLAST) or the PAM250 matrix (the standard matrix for FASTA and Smith-Waterman) was used. FASTA has several parameters that can be used to vary the speed and sensitivity of the search; normally FASTA ranks its output based on the initn score, which is based on locally similar regions without gaps. These regions can by found by looking for pairs of aligned identical amino acids (ktup=2) or single aligned identical amino acids (ktup=1). FASTA can also calculate for every sequence in the library an optimized scored in a 32 residue band around the best locally similar region; this substantially improves the performance of FASTA but slows it down only 2-fold (ktup=2/opt, ktup=1/opt). The analysis in Table 5 suggests that neither BLAST or FASTA (ktup=1 or ktup=2) perform as well as the more rigorous Smith-Waterman algorithm when searching for distantly related protein sequences. FASTA with ktup=2 performs significantly worse than BLAST, but there is no significant difference with ktup=1 (although FASTA is substantially slower). However, when all sequences are optimized, FASTA with ktup=2 or ktup=1 performs significantly better than BLAST. In fact, FASTA performs as well as the Smith-Waterman algorithm when an approximate similarity score in a 32-residue wide band is calculated for every sequence in the sequence database. Thus, although FASTA does not calculate a rigorously optimal similarity score, it performs as well as the Smith-Waterman algorithm when it calculates a very limited optimization.

Comparison of Scoring Matrices

The comparison method and statistical analysis described above is equally well-suited for evaluating the matrices and gap penalties used to calculate similarity scores. The PAM250 matrix, which is most widely used for amino acid sequence comparison, is based on data from 1976, when the protein sequence database contained only 77 267 residues in 767 sequences. This dataset was based entirely on direct protein sequencing; it would not be surprising to find that the far larger and less biased protein sequences that are available today from mRNA and DNA cloning and sequencing would generate a very different set of replacement costs. Jones et al. have described one such matrix [11] and note that there are some substantial differences between the 1978 PAM250 matrix and one based on 1991 data. For example, the most costly substitution in the 1978 data is a Cys -> Trp (-8); in the Jones matrix Cys -> Trp has a score of +1. Table 6 summarizes the performance of a variety of matrices of amino acids substitution scores, using the Smith-Waterman algorithm. Where not otherwise noted, the penalty for the first residue in a gap (insertion/deletion) was set to -12 and each additional residue in the gap decreased the score by 4. Two different versions of a scoring matrix based on the genetic code are shown; one ranges from +6 to -6 depending on whether 0 (+6), 1 (+2), 2 (-2), or 3 (-6) nucleotide changes are required; the other scores +4, +1, -2, and -6. Table 6 shows that, as expected,[9, 18] the PAM250 matrix performs significantly better than either the identity or matrices based on the genetic code.

The PAM250 matrix performs very well when compared with more modern matrices (Table 6). In fact, with the Smith- Waterman algorithm and gap penalties of -12/-4, the PAM250 matrix performs better than the PAM120 matrix, and those calculated by Gonnet et al.[12] and Jones et al.,[11] but with the exception noted below, the differences are not significant.

Table 5. Comparison of search algorithms

Algorithm	vs Smith-Waterman (pam250)				
	+	−	±1	Z	P
BLAST (120)	7	27	46	3.43	**0.0006**
BLAST (250)	5	30	47	4.23	**0.0001**
FASTA ktup=2(250)	1	33	44	5.49	$\mathbf{10^{-7}}$
ktup=2/opt	8	11	63	0.67	0.49
ktup=1	3	27	50	4.38	**0.0001**
ktup=1/opt	7	8	63	0.26	0.80
	vs BLAST (pam120)				
BLAST (250)	11	23	56	2.06	**0.04**
FASTA ktup=2 (250)	10	27	45	2.79	**0.005**
ktup=2 (opt)	24	8	50	2.83	**0.005**
ktup=1	15	17	47	0.35	0.72
ktup=1 (opt)	25	9	48	2.74	**0.006**

In their analysis of an exhaustive comparison of every sequence in a protein database with every other, Gonnet et al. suggest that the gap penalty reflect the evolutionary distance, but if that is not possible, a gap penalty of (-21/-2) would be superior to the more traditional (-12/-4) values. The Gonnet et al. matrix with -21/-2 gap penalties performs significantly worse than with more conventional gap penalties; the poor performance with a large initial gap penalty is consistent with the poor performance of BLAST, which does not allow gaps, and FASTA when gaps are not included. When a more conventional gap penalty is used, the Gonnet matrix performs as well as the other three matrices that are based on substitution data.

SUMMARY

We have developed a method for evaluating protein sequence comparison algorithms and the amino acid replacement scoring matrices used by these algorithms that is based on a simple statistical test. Comparison of the popular sequence searching algorithms suggest that the BLAST and FASTA (without optimization) algorithms perform worse than the Smith-Waterman algorithm but that FASTA performs as well as Smith-Waterman when a modest number of gaps are included in the similarity score. FASTA with ktup=2 and optimization runs about 25-times faster than the rigorous Smith-Waterman algorithm.

The PAM250 matrix, despite its age, performs surprisingly well compared to amino acid replacement matrices based on far more data. This is reassuring, but it is important to remember that we have focused on the task of identifying an unknown protein sequence by searching a protein sequence database; it seems likely that for different classes of proteins, more specifically tailored matrices may be more appropriate. In other experiments (data not shown), we have found that the performance of the Smith-Waterman algorithm and the FASTA algorithm is not affected significantly by the gap penalties in the range from -16/-2 to -8/-8. This is surprising, considering the strong sensitivity to gap penalties that have been described in comparisons of sequence and structure alignments. There is no question that alignments based on a constant penalty per residue (e.g. -8/-8) are often biologically unsatisfactory, but a constant gap penalty performs well for similarity searching. This may be a very useful insight, as it is considerably simpler to implement very rapid algorithms for constant gap penalties (Gene Myers, personal communication).

Table 6. PAM250 vs Alternative Scoring Matrices (Smith-Waterman algorithm)

matrix	+	-	±1	Z	P	ref.
Identity	3	39	35	55	10^{-8}	
Genetic code	1	44	31	6.41	10^{-9}	
Genetic code (+4/-6)	2	37	40	5.60	10^{-7}	
PAM120	11	16	55	0.96	0.34	5
Gonnet et al. (-21/-2)	6	16	63	2.13	**0.03**	12
Gonnet et al. (-16/-2)	8	12	60	0.89	0.37	12
Jones et al.	6	12	59	1.41	0.16	11
Henikoff	16	11	58	0.96	0.34	10

Our studies suggest that it is important to evaluate sequence comparison methods in the appropriate context. Algorithms that are more rigorous need not be better for identifying distantly related proteins. Scoring parameters may be based on more data, or a more sophisticated statistical analysis, and yet perform the same as ad hoc approaches. The most effective algorithm or scoring matrix need not be the same for both the similarity searching problem and the alignment problem; similarity searching can be evaluated with the methods described here, while alignments must be judged against multiple alignments or structural information.

It is unlikely that the dynamic-programming based algorithms that have been most successful for identifying protein sequences will be as effective for identifying other features in DNA sequences. The Needleman-Wunsch and Smith- Waterman dynamic programming algorithms, and their heuristic approximations, are effective for comparing distantly related protein sequences because they accurately model the evolutionary process of divergence from a common ancestor. Other biological features, e.g. regulatory sequences in DNA or splicing donor/acceptor sites, probably did not arise through a simple process of divergence. These features reflect convergence and other functional constraints; the best algorithms for detecting these features may be very different from Smith-Waterman, FASTA, or BLAST.

ACKNOWLEDGEMENTS

This research was supported by a research grant from the National Library of Medicine (LM04969). The use of the sign test for evaluating this data was suggested by Joe Felsenstein.

REFERENCES

1. S. Needleman and C. Wunsch. A general method applicable to the search for similarities in the amino acid sequences of two proteins. J. Mol. Biol. 48:444-453 (1970).
2. P. H. Sellers. On the theory and computation of evolutionary distances. *SIAM J. Appl. Math. 26:787-793 (1974).*
3. T. F. Smith and M. S. Waterman. Identification of common molecular subsequences. *J. Mol. Biol. 147:195-197 (1981).*
4. W. R. Pearson and D. J. Lipman. Improved tools for biological sequence comparison. *Proc. Natl. Acad. Sci. USA.85:2444-2448 (1988).*

5. S. F. Altschul, W. Gish, W. Miller, E. W. Myers, and D. J. Lipman. A basic local alignment search tool. *J. Mol. Biol. 215:403-410 (1990).*

6. O. Gotoh.An improved algorithm for matching biological sequences. *J. Mol. Biol. 162:705-708 (1982).*

7. D. J. Lipman and W. R. Pearson. Rapid and Sensitive Protein Similarity Searches. *Science. 227:1435-1441 (1985).*

8. R. F. Doolittle, D. F. Feng, M. S. Johnson, and M. A. McClure. Relationships of human protein sequences to those of other organisms. *Cold Spring Harb. Symp. Quant. Biol. 51:447-455 (1986).*

9. M. Dayhoff, R. M. Schwartz, and B. C. Orcutt, A model of evolutionary change in proteins, in Atlas of Protein Sequence and Structure, vol. 5, supplement 3, pp. 345-352.M. Dayhoff, ed., National Biomedical Research Foundation, Silver Spring, MD (1978).

10. S. Henikoff and J. G. Henikoff. Amino acid substitutions matrices from protein blocks. *Proc. Natl. Acad. Sci. USA. (in press).*

11. D. T. Jones, W. R. Taylor, and J. M. Thornton. The rapid generation of mutation data matrices from protein sequences. *Comp. Appl. Biosci. 8:275-282 (1992).*

12. G. H. Gonnet, M. A. Cohen, and S. A. Benner. Exhaustive matching of the entire protein sequence database. *Science. 256:1443-1445 (1992).*

13. W. R. Pearson. Identifying distantly related protein sequences. *Cur. Opinion in Struct. Biol. 1:321-326 (1991).*

14. M. S. Waterman and M. Eggert. A new algorithm for best subsequences alignment with application to tRNA-rRNA comparisons. *J. Mol. Biol. 197:723-728 (1987).*

15. W. R. Pearson. Searching Protein Sequence Libraries: Comparison of the Sensitivity and Selectivity of the Smith-Waterman and FASTA Algorithms. *Genomics. 11:635- 650 (1991).*

16. W. C. Barker, D. G. George, and L. T. Hunt, Protein Sequence Database, in Meth. Enz., vol. 183, pp. 31-49. R. F. Doolittle, ed., Academic Press, San Diego (1990).

17. A. S. Despande, D. S. Richards, and W. R. Pearson. A platform for biological sequence comparison on parallel computers. *CABIOS. 7:237-247 (1991).*

18. D. F. Feng, M. S. Johnson, and R. F. Doolittle. Aligning amino acid sequences: comparsion of commonly used methods. *J. Mol. Evol. 21:112-125 (1985).*

ALGORITHMIC ADVANCES FOR
SEARCHING BIOSEQUENCE DATABASES[*]

Eugene W. Myers

Department of Computer Science
University of Arizona
Tucson, AZ 85721

INTRODUCTION

The similarity search problem can loosely be phrased as follows. Given a *query* sequence of length, say P, a *database* of a total of, say N, characters of sequence data, and a *threshold D*: find all "database entries similar within threshold" to the query. The objective placed between quotes can and has been interpreted in a variety of ways. In the case of similarity searches over biosequence databases, what is usually touted as the most sensitive standard is to find local alignments between substrings of the query and a database entry whose similarity measure under, say, Dayhoff-matrix substitutions [1] and affine gap costs, is greater than D. As it is currently realized with the Smith & Waterman [2] dynamic programming algorithm, such a search is generally felt to be too computationally intensive to be routinely performed.

At the current time, the size N of biosequence databases is growing at an exponential rate, and many investigators foresee a time when there will be terabytes of sequence information available. On the other hand, the size of a query, P, tends to remain fixed, e.g. a protein sequence averages about 300 residues and the longest are about 1000 residues. So designers of efficient computational methods should be principally concerned with how the time to perform a search grows as a function of N. Despite this all the currently used methods take an amount of time that grows linearly in N, i.e. they are $O(N)$ algorithms. This includes not only rigorous methods like Smith & Waterman, but also popular heuristics, e.g., FASTA [3] and BLASTA [4], and even the systolic array chips, e.g., FDF [5] and BISP [6]. What this implies is that when a database increases in size by a factor of 1000, then all these methods will be taking 1000 times longer to search that database. The only difference among them is their coefficients of proportionality, e.g. Smith & Waterman takes roughly $1.0N$ milli-seconds, BLASTA roughly $2.0N$ micro-seconds, and the BISP chip roughly $0.3N$ micro-seconds for a search against a typical protein query sequence on a typical workstation. Thus while a custom chip may take about 3 seconds to search 10 million residues or nucleotides, it will take 3000 seconds, or about 50 minutes when there are 10 billion symbols to be searched. And this is the fastest of the linear methods: BLASTA

[*]This work was supported in part by NLM Grant R01 LM04960, NSF Grant CCR-9001619, and the Aspen Center for Physics

will be taking hours and Smith & Waterman months on these future databases. One could resort to parallelism, but such machinery is beyond the budget of most investigators, and it is unlikely that speedups due to hardware technology improvements will keep up with sequencing rates in the next decade.

What would be very desirable, and as argued above probably essential, is to have search methods whose running time is sublinear in N, i.e., $O(N^\alpha)$ for some $\alpha < 1$. For example, suppose there is an algorithm that takes $O(N^{.5})$ time, which is to say that as N grows, the time taken grows as the square root of N. For example, if the algorithm takes about 10 seconds on a 10 million symbol database, then on 10 billion symbols, it will take about $\sqrt{1000} \approx 31$ times longer, or about 5 minutes. Note that while it was slower than the hypothetical chip on 10 million symbols, it is faster on 10 billion. Figure 1 illustrates this "crossover": the size of N at which the $O(N^{.5})$ algorithm overtakes the $O(N)$ method. Similarly, an $O(N^{.2})$ algorithm that takes, say, 15 seconds on 10 million symbols, will take about one minute or only 4 times longer on 10 billion. To forcefully illustrate the point, we chose to let our examples be slower at $N = 10$ million than the competing $O(N)$ chip. As will be seen later the sublinear algorithm to be described is actually already much faster on database of this size, having crossed over at a much smaller value of N. The other important thing to note is that we are not considering heuristics. What we desire is nothing less than algorithms that accomplish exactly the same computational task as Smith & Waterman or some other dynamic programming method but is much faster because the computation is performed in a clever way.

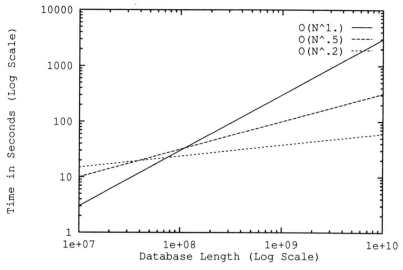

Figure 1. Sublinear Versus Linear Methods

Having now described the challenge, the remainder of the paper turns to a presentation of recent results by theoretical computer scientists that portend the possibility of an affirmative answer. Computer scientists have long focused research efforts on a simplification of the general problem, called approximate keyword matching, where one scores an alignment simply by its number of differences (mismatches, insertions, and deletions) and where one is constrained to match the entire query against a substring of the database. That is, scoring is restricted to "unit costs" as opposed to arbitrary weightings (e.g. PAM scores and gap costs), and alignments must globally align the query (but not the database entry). Thus this is a "toy" problem from the view of applications in molecular biology. But when faced with the daunting task of designing a sublinear algorithm, such simplifications often help to focus one's thinking and obtain preliminary results. Indeed, a provably sublinear algorithm for approximate keyword matching has been designed recently by this author and it is sketched later in this paper. It is worth noting that it is highly efficient compared to previous algorithms: for relatively stringent matches it is 3 to 4 orders of magnitude more efficient than the equivalent dynamic programming computation on a database of one million characters. As the paper winds it way through various recent and highly efficient algorithms for approximate keyword matching, we will digress to note how each method might extend to the general problem which is a focus of current research for the author.

PRELIMINARIES AND HISTORY

Given a relatively short query sequence W of length P, a long "database" sequence A of length N, and a threshold D, the *approximate keyword search problem* is to find all substrings of A that align with W with not more than D insertions, deletions, and mismatches. Thinking of insertions, deletions, and mismatches as editing operations that transform one sequence to another, one defines the *simple Levenshtein distance* [7], $\delta(V, W)$, as the minimum number of such operations needed to transform V into W. For example, if $V = ababaa$ and $W = abbba$ then

ababaa	ababaa	ababaa
abbba-	ab-bba	-abbba

are alignments between them. The first two alignments both realize the edit distance, 2, between V and W, and the third is non-optimal as it codifies one deletion and three substitutions. Terming the Levenshtein measure a distance is justified by the fact that it forms a metric space over the domain of sequences. Let $A[i, j]$ denote the substring of A consisting of its ith through j^{th} characters. The objective of the approximate keyword search problem can now be formally stated as finding all pairs i, j such that $\delta(W, A[i, j]) \leq D$. For such a problem, we say that the maximum *mismatch ratio* is $\varepsilon = D / P$ and that we are searching A for ε-*matches* to W. That is, ε is just the search threshold rephrased in length relative terms. It is the fraction of errors permitted per unit of query length. A 10%-match to a query of 50 characters must involve 5 or less edits.

Throughout the paper, the approximate keyword problem will be the primary focus. The ultimate objective, however, is to design an efficient algorithm for finding *local* alignments under a *generalized* Levenshtein measure [7]. Rather than scoring alignments simply by counting edit operations, a user is free to specify a cost, $s(a, b)$, for substituting a for b, another, $d(a)$ for deleting a, and yet another, $i(b)$ for inserting b. The score of an alignment is then considered to be the sum of the costs assigned to the edit operations of the alignment, and the generalized Levenshtein measure $\delta(V, W)$ is the cost of a minimal cost alignment. This general measure is not necessarily a metric but the conditions for it to be so are well understood [8]. The other additional freedom, is that now one is looking for any substring of the query that aligns well with a substring of the database and is sufficiently long to be statistically rare. Formally, the *weighted, local-alignments search problem* is: Given query W, database A, threshold D, minimum substring length M, and costs $s(a, b)$, $d(a)$, and $i(b)$, find all quadruples $g \leq h$, and $i \leq j - (M-1)$ such that $\delta(W[g, h], A[i, j]) \leq D$. This formal characterization is only one of several forms in which the problem may be presented. For example, in those regions where there are several overlapping pairs of substrings above threshold, then one might want to report only the one that is, in some sense, locally dominant. Another variation requires one to maximize alignment score rather than minimize, but transforming one to the other simply requires multiplying all basic costs by -1. Such subtleties are unnecessarily distracting here, the essence of the generalization required in going from the approximate keyword search problem to the biologically relevant search problem are (1) to allow arbitrary gap and substitution weights, and (2) to consider substrings of the query.

The simple approximate keyword search problem has been much studied. Sellers [8] presented the obvious $O(PN)$ algorithm as a slight variation of the classic dynamic programming algorithm for the sequence versus sequence comparison problem. This variation, another allowing dangling ends at no-cost, and the Smith & Waterman algorithm are all simply variations of the dynamic programming recurrence in which the boundary conditions are modified appropriate to the problem. The next development was the independent reporting by both Ukkonen [9] and Mount & Myers [10] of a practical and simple $O(DN)$ expected time algorithm. Not long thereafter, Landau & Vishkin [11] and Myers [12] arrived at two different algorithms, that both achieved $O(DN)$ worst case time but required $O(N)$ space. More recently, Galil & Park [13] and Ukkonen & Wood [14] have designed distinctive $O(DN)$ worst-case time algorithms that require only $O(P^2)$ space. In another direction, Wu, Manber, & Myers [15] applied the 4-Russians paradigm to arrive at an $O(DN/\log P)$ expected-time algorithm requiring only $O(P)$ space. Unfortunately, all these developments take special advantage of the unit-cost feature of the simple Levenshtein measure, and so do not have the potential to provide speedups for the general problem.

In the last two or three years, a number of distinct lines of continued development on the approximate keyword search problem have lead to sublinear algorithms, and there is at least the possibility that these contributions can impact the general problem. All the methods assume that the sequences are over a finite alphabet Σ of size $S = |\Sigma|$. For example, S is 4 for DNA sequences, 20 for protein sequences, and 128 for ASCII texts. This paper discusses two of the three recent developments and how they may be extended to the general problem.

In one direction, a number of investigators have designed methods that employ a "prefilter" to quickly eliminate regions of the database that cannot possible match the query within the given threshold. In approximately the same time frame, Ukkonen [16] designed a prefilter based on the *k-tuple measure* which can be customized to provide an $O(N)$ expected-time algorithm. Chang and Lawler [17] used the *spanning measure* of Ehrenfeucht and Haussler [18], to design an even faster $O(DN \log P / P)$ expected-time algorithm. Recently Wu and Manber [19] have designed a variant of the Chang and Lawler method that while not asymptotically superior is very efficient in practice. All these algorithms require that D be less than $P/(\log P + O(1))$ or the complexity claims do not hold as the prefilter does not eliminate enough of the database. Chang and Lawler term their algorithm sublinear in the sense of Boyer and Moore's [20] exact matching algorithm because when D is $O(P/\log P)$ their algorithm examines only a fraction of the characters in the database. Note however, that it is not sublinear in the sense used in the introduction because the algorithm still takes time linear in N. One of the other drawbacks of these methods are that as P gets larger the stringency of a match must be tightened as the constraint on D implies that ε must be less than $1/(\log P + O(1))$. The section on prefilter algorithms concludes with an extension of these methods to the general local alignments problem, but argues that the postulated design is probably not effective in practice.

In a second direction, this author has designed an algorithm that *given a precomputed index* of the database A, finds rare approximate matches in time that is sublinear in N in the sense of the introduction, i.e., takes time proportional to N^α for some $\alpha < 1$. More precisely, the algorithm requires $O(DN^{pow(\varepsilon)}\log N)$ expected-time where (1) $pow(\varepsilon)$ is increasing and concave, (2) $pow(0)=0$, and (3) $pow(\varepsilon)$ is dependent on alphabet size S. The algorithm is superior to the $O(DN)$ algorithms when ε is small enough to guarantee that $pow(\varepsilon) < 1$. For example $pow(\varepsilon)$ is less than one for $\varepsilon < 33\%$ when $S = 4$ (DNA alphabet), and for $\varepsilon < 56\%$ when $S = 20$ (Protein alphabet). Figure 2 later in the paper precisely plots the function pow for these two choices of S. Note that this algorithm has an advantage over the prefilter algorithms in that the degree of sublinearity just depends on ε and not on P. That is, the prefilter algorithms require that ε be less than $O(1/\log P)$ which becomes smaller as P grows. On the other hand, we require a precomputed $O(N)$ space index structure, whereas the prefilters are purely scanning based, requiring only $O(P)$ working storage. However, our algorithm is effective for a much wider range of ε and is orders of magnitude more efficient than all other methods for small ε. As a consequence it portends the possibility of an effective algorithm for the general local alignments problem. We give some preliminary experimental results at the conclusion of the paper.

Finally, we note the work of Gonnet et al. [21] for which we have still not seen a technical manuscript. Their approach also uses a precomputed index and was designed directly for the general local alignments problem. They have reported execution times that are significant faster than those of the Smith & Waterman algorithm. It appears that they have not been able to demonstrate mathematically the degree of sublinearity of their algorithm, but have estimated on the basis of empirical evidence that their algorithm takes time proportional to about $N^{.6-.8}$ for the all-against-all comparison problem. While we believe that we understand their method, without a formal technical description we think it unwise to describe what we can only surmise.

PREFILTER ALGORITHMS

The basic structure of all the prefiltering algorithms described in this section is as follows. In a scan of the database, quickly eliminate regions that can't possibly match via some easily computed criterion. For any region not eliminated, apply a more expensive dynamic programming computation to determine if there is a match. The algorithms vary in the manner in which they eliminate regions from consideration. However, all are dependent on the threshold D in that the greater D becomes the less chance a region has of being eliminated. Thus in order for these algorithms to be efficient, D must be small enough that the time spent checking unfiltered regions for matches via dynamic programming is in expectation less than the time for the scan. In all three of the unit cost algorithms below, the limit for D is $O(P/\log P)$, i.e., ε must be $O(1/\log P)$.

K-Tuple Measure

An algorithm by Ukkonen [16] centers on using the *k-tuple measure* between strings to serve as a lower bound on their simple Levenshtein distance. For any $k > 0$, let the k-tuple distance between sequences W and V be defined as:

$$\delta^k(W,V) = \sum_{x \in \Sigma^k} |\#_W(x) - \#_V(x)|$$

where $\#_W(x)$ is the number of occurrences of k-tuple x in W. From one point of view, δ^k is the Manhattan distance between the profile vectors of V and W in their underlying S^k-dimensional space*. As an example, $\delta^2(abba, aaba) = |\#_W(aa) - \#_V(aa)| + |\#_W(ab) - \#_V(ab)| + |\#_W(ba) - \#_V(ba)| + |\#_W(bb) - \#_V(bb)| = |0-1| + |1-1| + |1-1| + |1-0| = 2$. The essence of the way this measure is used to eliminate regions is encapsulated in the following lemma.

Lemma: For all k, if $\delta^k(W, A[i-(P-1), i]) > 2kD$ then the best approximate match of W to a substring of A ending at position i, has score greater than D.

This lemma is quit easy to prove, the basic insight needed is that each difference in an alignment between W and the substring of A creates at least k k-tuples of W that are not in the substring of A.

---

*The k-tuple profile of a string W is $< \#_W(x) >_{x \in \Sigma^k}$, and the Manhattan distance between n-dimensional points $(x_1, x_2, \cdots x_n)$ and $(y_1, y_2, \cdots y_n)$ is $\sum_{i=1}^{n} |x_i - y_i|$.

To lever the lemma as a prefilter simply requires that one compute $Val(i) = \delta^k(W, A[i-(P-1), i])$ at each index i of A in a left to right scan of increasing i. One then checks if $Val(i) \leq 2kD$ and for those i for which it is, one then invokes a traditional $O(DP)$ dynamic programming algorithm to check for an approximate match within threshold D. If one chooses $k = \log_S P + 1$, it can be shown that the total amount of time spent on calls to the dynamic programming subroutine is less than $O(N)$ in expectation, provided that $D = O(P/\log P)$. Moreover, for this choice of k the scan is easily performed in $O(N)$ time using an $O(PS)$ table $Cnt[x]$, where x ranges over all $S^k = PS$ k-tuples. The trick to doing so, is to encode the k-tuples as integers in the range $[0, PS-1]$ in the natural way, and to maintain Cnt so that for all x, $Cnt[x] = \#_W(x) - \#_{A[i-(P-1), i]}(x)$ after completing the computation of $Val(i)$. Computing $Val(i)$ from $Val(i-1)$ simply amounts to observing the effect on the k-tuple metric of the removal of the tuple $A[i-P, i-P+(S-1)]$ and the introduction of the tuple $A[i-(S-1), i]$ on the Cnt table. In summary, for $D = O(P/\log P)$ Ukkonen's algorithm takes $O(N)$ time in the expected case.

Maximal Span Measure

Another simple concept of sequence similarity that can be used to bound simple Levenshtein matches was first introduced by Ehrenfeucht and Haussler [18]. The *maximal span measure* of V with respect to W, $\delta^W(V)$ is the minimum number of positions whose removal from V leaves intervening substrings between them that are all substrings of W. It is not hard to show that this measure can be computed by greedily, i.e., find the longest prefix of V that matches a substring of W, remove this prefix and the character following it, and then reiterate the process on what's left of V. The number of characters deleted after each common prefix is $\delta^W(V)$. For example, if $V = aababaabbbb$ and $W = abbab\text{-}baab$ then, the procedure just sketched partitions V as $\overline{aababaabbbb}$ implying that $\delta^W(V)$ is 2.

Chang and Lawler [17] used this measure to eliminate regions of the database from consideration without necessarily having to scan every character of the database. Thus the algorithm is sublinear in the sense of Boyer and Moore's [20] exact string matching algorithm, but not in the sense of taking time $O(N^\alpha)$ for $\alpha < 1$. They proceed by partitioning the database into $O(N/P)$ panels each of size $(P-D)/2$. Since edit operations are unit cost, if W approximately matches a substring of A with D-or-less differences, then the substring of A matched must be of length at least $P-D$ since one can delete at most D characters of W in a given match. But then it follows that such a substring must completely span one of the panels. With this observation, it follows that a given panel, $pan = A[i, i+(P-D)/2-1]$, can be eliminated if $\delta^W(pan) > D$ because $\delta^W(pan)$ bounds the number of edit operations in a best alignment between pan and any substring of W. As long as D is sufficiently small, specifically, $O(P/\log P)$, then it can be shown that the expected total time spent spent doing dynamic programming to check panels that are not eliminated is less than the time to do the scan described in more detail below.

To eliminate a panel, $A[i, i+(P-D)/2-1]$ it suffices to show that the spanning measure of the panel with respect to the query W is greater than D. Let $Match(p) = \max\{l : \exists j\ W[j, j+l] = A[p, p+l]\}$, that is, the length of the longest match between a substring of W and a substring of A beginning at position p. Proceeding greedily, set $S_0 = i$, and then for increasing values of d from 1 up to $D+1$, compute $S_d = S_{d-1} + Match(S_{d-1}) + 2$. This recurrence captures the greedy computation of the maximum span measure of the panel. If $S_{D+1} < i+(P-D)/2$ then the measure of the panel is greater than D and it can be eliminated. On the other hand if at any point during the computation of S_{D+1}, a d is reached for which $S_d \geq i+(P-D)/2$ then one can quit prematurely and move immediately to a dynamic programming check of the panel. Note that unlike Ukkonen's algorithm, all that is known is that there might be a match involving the panel (as opposed to ending at some specific position) and so one must check W against all of $A[i-(P+3D)/2, i+(P+D)-1]$.

Using a suffix tree [22,23] or suffix automaton [24] of the query W, one can compute $Match(p)$ in time proportional to its value. In expectation, $Match(p)$ is $\log_S P$ and so the time spent scanning a panel is $O(D \log P)$. This gives a total expected time complexity for the scan phase of $O(DN \log P / P)$. Note that since $D = O(P / \log P)$ by assumption, it follows that the algorithm is always $O(N)$ and is generally superior in expectation to Ukkonen's algorithm as it generally scans only a portion of each panel.

Recently, Wu and Manber [19] non-asymptotically improved upon the maximal span measure prefilter by combining it with the k-tuple approach of Ukkonen as follows. They chose $k = \log_S P + 1$ and rather than computing $Match(p)$ with a suffix tree they estimate it by checking if $A[p, p+(k-1)]$, $A[p+1, p+1+(k-1)]$, $\cdots$ match a k-tuple in W using a precomputed PS-element, tuple-indexed array of booleans whose true entries are exactly those whose index is a substring of W. $Match(p)$ is estimated to be the right-end of the first k-tuple that doesn't match one in W. Certainly this procedure computes either $Match(p)$ or a larger value. In expectation, it is a tight estimate of $Match(p)$ and the resulting algorithm is significantly more efficient in practice than one using a suffix tree.

Generalizing to Weighted Local-Alignment Searches

So now we ask if the methods above, specifically designed for the unit cost model of comparison can be extended to the problem of finding local similarities between the query and database of score less than threshold D and length at least M, under a general Levenshtein scoring scheme. The answer is positive and is an extension of Ukkonen's k-tuple method requiring significant modification.

For every k-tuple, x, precompute $Best[x] = \min\{\delta(x,y) : y$ is a substring of $W\}$. That is, $Best[x]$ is the best scoring alignment between x and some substring of W. We will also need arrays $Sufx[x] = \min\{\delta(z,y) : y$ is a substring of W and z is a suffix of $x\}$, and $Remx[x] = \min\{\delta(z,y) : y$ is a substring of W and z is a suffix of x of length at least $k - M\ (mod\ k)\}$. These two arrays account for aligned portions of the database that may not involve an entire k-tuple. Such tables can be computed in $O(S^k P)$ time by incrementally computing values over the complete trie of tuples. Thus the precomputation amounts in complexity to performing a dynamic programming comparison of W against a database of length S^k.

Given the tables, our bound on local matches ending at position i, $Bound(i)$, is as follows:

$$Bound(i) = \min_{t \geq m}\{\ (\text{ if } t = m \text{ then } Remx[A[i \blacklozenge t]] \text{ else } Sufx[A[i \blacklozenge t]]) + \sum_{j=1}^{t-1} Best[A[i \blacklozenge j]]\ \}$$

where $m = \lceil M/k \rceil$ and $A[i \blacklozenge t]$ is the k-tuple $A[i-tk+1, i-(t-1)k]$ of the database. Intuitively, the bound is the sum of the best possible alignments of consecutive k-tuples ending at i, minimized over all possible stretches of the database ending at i and of length at least M. Certainly $Bound(i)$ is not greater than the score of the best possible alignment of a *substring* of the query against a substring of A ending at position i because we are ignoring the ordering requirement of sequence alignments. $Bound$ values are easily computed in an $O(N)$ time, $O(k)$ space scan of A using the easily verified tandem recurrences: $Sum(i) = Sum(i-k) - Best[A[i \blacklozenge m]] + Best[A[i \blacklozenge 1]]$ and $Bound(i) = \min\{Sum(i) + Remx[A[i \blacklozenge m]], Sum(i) + Best[A[i \blacklozenge m] + Sufx[A[i \blacklozenge m+1]], Bound(i-k) + Best[A[i \blacklozenge 1]]\}$. Only those regions for which $Bound(i)$ is not greater than the threshold D need be checked for local substring matches with a dynamic programming algorithm.

It will frequently be the case that when $Bound(i) \leq D$ for a given i it will also be D-or-less for an interval of indices about i. Thus in practice, one will want to accumulate such an interval and then run a Smith & Waterman style dynamic programming algorithm to determine if there is local alignment that involves a substring of A ending at an index in the given interval. Moreover, if desired one can bound the length of all substrings of A that could match a substring of P. Specifically, the longest possible matching stretch is given by:

$$P + \left\lfloor \frac{D - (\sum_{j=1}^{P} \min_{a \in \Sigma} s(W[j], a))}{\min_{a \in \Sigma} d(a)} \right\rfloor$$

where $s(a, b)$ is the score of aligning a with b and each deleted character a is penalized $d(a)$. This bound can be used to further refine $Bound(i)$ and to limit the range of the dynamic programming computation in an *a priori* fashion.

Only experiments will determine if the above concept works well in practice. It is estimated that it will be effective only for short query sequences and relatively stringent threshold values. The reasoning behind these prediction is as follows. The unit cost algorithms work only for a limited range of D (or, equivalently, ε). In practice ε has to be less than about $\frac{1}{2}\log_S P$. For example, for DNA sequences and a query of length 100, the methods fail to eliminate enough positions when ε becomes only 7% or larger. As the query gets larger, this break-even percentage becomes ever smaller. For protein sequences ($S = 20$) the break-even percentage for ε is higher, about 30%. Nonetheless, if one thinks about what this translates into as a thresholded search under Dayhoff-matrix scores, one guesses that this corresponds to a very strong threshold. Moreover, the longer the sequence gets, the worse performance will become. Intuitively, the longer the sequence gets the more chance there is that $Best[x]$ becomes negative for every x. Unless the preponderance of $Best$ values are positive, i.e. $\Sigma_x Best[x]/S^k > 0$, the prefilter postulated in this section will not be effective.

A SUBLINEAR APPROXIMATE MATCH ALGORITHM

We now examine the structure of a sublinear algorithm for approximate matching designed by this author [25]. It has the feature of being truly sublinear for a much wider range of ε than the prefilter algorithms, as well as being more efficient in practice. Thus it portends a possible extension to the general similarity search problem that will significantly improve on the time taken by the Smith & Waterman algorithm while performing the equivalent computation.

A Sketch of the Algorithm

The algorithm depends on two critical ideas. The first is the use of an index of the database to rapidly find all instances of a given sequence in the database. It is clear that in order to be sublinear, some kind of precomputed data structure encoding information about the database is required. Without such, one must at a minimum spend $O(N)$ time reading the database (or an $O(N)$ portion in the case of the Chang and Lawler prefilter). Index data structures for finding all substrings of the database A that exactly match the query have been much studied. The asymptotically most efficient are the suffix tree [22,23] and suffix array [26] data structures. A suffix tree can be built in $O(N \log S)$ time, occupies $O(N)$ space, and can find all occurrences of a query of length P in $O(P \log S + H)$ time where H is the number of occurrences reported. If one permits the suffix tree to occupy $O(NS)$ space, then it can be built in $O(N)$ time and queries take $O(P + H)$ time. Suffix arrays are a particularly space efficient alternative to suffix trees, requiring only $2N$ integers, but they take $O(N \log N)$ time to build and searches take $O(P + H + \log N)$ time. However, for the algorithm ultimately developed here, all queries will be of length $k = \log_S N$. Thus, each query can be uniquely encoded as an $O(N)$ integer in the usual way, and an index that is a store of the results of all possible queries (i.e., lists of indices where the corresponding strings of size k appear in A) in an $O(N)$ table. This particularly simple index structure can be built in $O(N)$ time, occupies $2N$ integers of space, and performs queries in $O(P+H)$ expected-time.

The second critical idea is that of the *neighborhood* of a sequence. The D-*neighborhood* of a sequence W is the set of all sequences that can be aligned to W with less than D differences, i.e., $N_D(W) = \{V : \delta(V, W) \leq D\}$. For example, the 0-neighborhood of *abbaa*, $N_0(abbaa) = \{abbaa\}$ as *abbaa* is the only string that can be aligned to *abbaa* with 0 differences. However, the 1-neighborhood (with respect to the alphabet $\Sigma = \{a, b\}$) contains 15 sequences: $N_1(abbaa) = \{$ *aabaa*, *aabbaa*, *abaa*, *abaaa*, *ababaa*, *abba*, *abbaa*, *abbaaa*, *abbaab*, *abbaba*, *abbba*, *abbbaa*, *babbaa*, *bbaa*, *bbbaa* $\}$.

Now the simple way of combining these two ideas in order to find all approximate matches to W with D-or-less differences is to simply generate all the sequences that align to W with D-or-less differences, i.e. $N_D(W)$, and to then efficiently find the, say, left end of every substring of A that exactly matches one of these sequences using an index. This takes roughly $O(PZ+H)$ time where Z is the size of the D-neighborhood and H is the number of matches found. There is a small and a large problem with this approach. The small problem is that a given left end may be reported more than once. For example, for $N_1(abbaa)$, wherever one finds *abaaa* one will also find *abaa* and so report the left end of occurrences of *abaaa* twice. This problem is easily rectified, by instead generating the *condensed D-neighborhood* of W which is the set of all strings in the D-neighborhood of W that do not have a prefix in the neighborhood, i.e., $\overline{N_D}(W) = \{V : V$ in $N_D(W)$ and no prefix of V is in $N_D(W)\}$. For example, $\overline{N_1}(abbaa) = \{$ *aabaa*, *aabbaa*, *abaa*, *ababaa*, *abba*, *abbba*, *babbaa*, *bbaa*, *bbbaa* $\}$ contains *abaa* but not *abaaa*. By its construction, one will never report a given position more than once and in fact the procedure is made a bit more efficient as condensed neighborhoods are always smaller than the corresponding neighborhood.

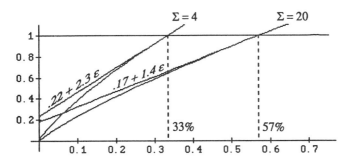

Figure 2. Plots of *pow(ε)* and Bounding Tangents

Now the *big* problem is that this simple algorithm takes time that is a function of Z and a little thought reveals that Z grows explosively in both D and P, even for condensed neighborhoods. For example, a quick "back of the envelope" calculation (over)estimates Z as growing like $(2SP)^D/D!$. Thus this approach is only viable for "sufficiently small" queries and thresholds. For our purposes, "sufficiently small" is when $P = \log_S N$, for then one can rigorously prove [25] that Z and the expected value of H are not greater than $N^{pow(\varepsilon)}$ where $pow(\varepsilon) = \log_S (c+1)/(c-1) + \varepsilon \log_S c + \varepsilon$ and $c = \varepsilon^{-1} + \sqrt{1+\varepsilon^{-2}}$. The function $pow(\varepsilon)$ is a bit hard to understand analytically but Figure 2 above makes it clear that it is a concave function of ε that starts at the origin and rises slowly (relative to the prefilter algorithms) to 1 as ε increases. The larger the alphabet size, S, the slower the curve rises. The figure shows the curve for the DNA alphabet ($S = 4$) and the protein alphabet ($S = 20$). $Pow(\varepsilon)$ is less than 1 for $\varepsilon \leq 33\%$ for DNA and $\varepsilon \leq 56\%$ for proteins. Furthermore, for a quick approximation one can use the bounding tangent lines $.22 + 2.3\varepsilon$ and $.17 + 1.4\varepsilon$ for DNA and proteins, respectively. In [25] a cleverly constructed algorithm* is given that generates a condensed D-neighborhood in $O(DZ)$ worst-case time and further looks up all occurrences in the database using the simple index structure in an additional $O(DH)$ expected time. Thus, for small queries of length $L = \log_S N$, the expected time taken by the simple approach is $O(DN^{pow(\varepsilon)})$.

So in order to overcome the central limitation, we need to extend an efficient algorithm for queries of length L into one that efficiently handles longer queries. The basic idea is to break W into pieces of length L, find ε-matches to these using the simple approach, and then to try extending these to ε-matches to encompass all of W. This basic idea has been incarnated in various heuristic forms, but in order to maintain rigor one must be very careful about how the decomposition and extension steps are performed.

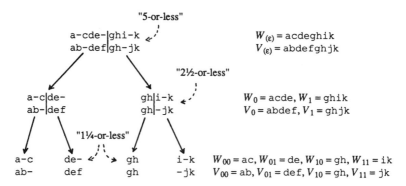

Figure 3. Illustration for Division Lemma

Consider dividing the query W of length P in a binary fashion until all pieces are of size L. Assume for simplicity that P is a power of 2 times L so that the division process results in pieces of exactly size L. So that we may specify various pieces formally, let W_α for labels $\alpha \in \{0, 1\}^*$ be recursively defined by the equations: $W_\varepsilon = W$, $W_{\alpha 0} =$ first_half_of(W_α), and $W_{\alpha 1} =$ second_half_of(W_α). The crucial lemma below follows from a simple application of the Pigeon-Hole Principle.

Lemma: If W ε-matches V, then there exists a label α of a length L piece of W, such that for every prefix β of α, W_β ε-matches a substring, say V_β, of V. Moreover, $V_{\beta a}$ is a prefix or a suffix of V_β according to whether a is 0 or 1.

Figure 3 depicts the decomposition of a query W and illustrates the proof of the lemma. If W ε-matches a string V, then this implies there is a "D-or-less" alignment between them. Consider bisecting the alignment at any column that separates W_0 and W_1 (there are two

---

*The algorithm involves computing rows of a dynamic programming matrix in response to a backtracking search that essentially traces a trie of the words in the condensed D-neighborhood of W.

choices in the example of Figure 3). By the Pigeon-Hole Principle, one of these two portions must be a "$D/2$-or-less" alignment, for if both were "more-than-$D/2$" alignments, then the original would have had to have been a "more-than-D" alignment. Suppose for illustration it was W_1 as in the figure. Then W_1 must ε-match some suffix of V as $(D/2)/(P/2) = \varepsilon$. Recursively applying the argument above to W_1, there must be a "$D/4$-or-less" alignment between either W_{10} or W_{11} with a prefix or suffix of V_1, respectively. Thus the lemma follows: there is some sequence of halves leading to a terminal L-sized piece such that the progressively smaller pieces all ε-match some progressively smaller parts of V.

The lemma can be used to construct an "air-tight" extension strategy for extending matches to smaller pieces into matches to larger pieces as follows. First find ε-matches to the P/L smallest pieces of length L using the simple algorithm. Suppose for example, that one of these pieces is W_{1010} and one discovers that it matches $A[i..j]$. Then see if this match can be extended to an ε-match of W_{101} and some substring of A whose left end is at position i. Do this by running a standard dynamic programming algorithm on the relevant regions of the two strings. If the attempted extension fails then one need not proceed further. But if the extension is successful, say W_{101} is matched to $A[i..k]$ then try ε-matching W_{10} against a string whose *right* end is at position k, and so on until either one fails at some step, or one ultimately discovers an ε-match to all of W. By the lemma, no such match to W can be missed because there must be some sequence of small to larger pieces for which one obtains an ε-match at every level.

Efficiency Analysis and Empirical Results

The first stage of the sublinear algorithm finds all matching sequences in the (εL)-neighborhood of each of the P/L pieces of size L into which the query is partitioned. Doing so takes $O((P/L)(\varepsilon L)N^{pow(\varepsilon)}) = O(\varepsilon P N^{pow(\varepsilon)}) = O(DN^{pow(\varepsilon)})$ time and delivers $O(PN^{pow(\varepsilon)}/L)$ matches in expectation by the characterization of the quantities Z (neighborhood size) and H (hit ratio) presented earlier. For the second, extension phase, the critical part of the time complexity analysis is the observation that while the query substrings and threshold both double with each successive level, it is also the case that the expected number of matches surviving at each level is dropping hyperexponentially (*except* where the search will reveal a distance D match to W). That is, while the time spent in dynamic programming to check an extension increases by a factor of 4 at each level, this is overwhelmed by the fact that the number of random matches surviving to the kth level decreases hyperexponentially as $N^{-2^k(1-pow(\varepsilon))+1}$. Thus the time spent in extensions is dominated by the first level of the extension phase and this requires checking $O(PN^{pow(\varepsilon)}/L)$ length L matches with $O(\varepsilon L^2)$ time dynamic programming extension checks for a total of $O(\varepsilon P N^{pow(\varepsilon)}L) = O(DN^{pow(\varepsilon)}\log N)$ expected time. Because this last cost dominates all others, we conclude that the total time consumed in expectation is $O(DN^{pow(\varepsilon)}\log N)$ under the assumption that the database is the result of random Bernouilli trials. However, if the database is not random but preconditioned to have H matches to W, then a total, per match, of $O(DP)$ additional time will be spent in dynamic programming during the extension of seed matches into one encompassing all of W. Our algorithm's complexity in this case is $O(DN^{pow(\varepsilon)}\log N + HDP)$, i.e., the algorithm takes $O(DN^{pow(\varepsilon)}\log N)$ time in expectation to eliminate the portions of the database not containing matches to W, and does one $O(DP)$ dynamic programming computation per matching region. Consequently, our algorithm is not an improvement of the $O(N+D^2)$ algorithm [27] for the sequence-versus-sequence problem.

Nonetheless, for searching problems where the database is large and ''sufficiently'' random, this algorithm can find near matches with great efficiency and no loss in sensitivity. This is illustrated in Figure 4 which plots the times for an implementation of our algorithm over a one million nucleotide DNA sequence and a four million residue protein sequence against a query of length 80. We compared our implementation against an implementation of the standard $O(NP)$ dynamic programming algorithm [8], the $O(DN)$ expected-time algorithm of Ukkonen [9], and Mount & Myers [10], and a novel use of the 4-Russians paradigm that permits the dynamic programming matrix to be computed 5 entries at a step [15]. In all cases the software had been written at an earlier time by this author and represents his best efforts at efficient code. All experiments were performed on a SparcStation 2 with 64 megabytes of memory and all code was compiled under the standard SunOS C-compiler with the optimization option on. The plotted timing curves in Figure 4 are labeled as follows: T_{DP} for the standard dynamic programming algorithm, T_U for the Ukkonen algorithm, T_{4R} for the 4-Russians algorithm, and T_{slam} for our sublinear algorithm. A logarithmic time scale is used because the sublinear algorithm's time performance increases exponentially in D. Thus the curve for T_U is shaped like a log curve because it is actually a straight line on a normal scale. T_{DP} and T_{4R} are straight lines because the complexity of their underlying algorithms depend only on P and N.

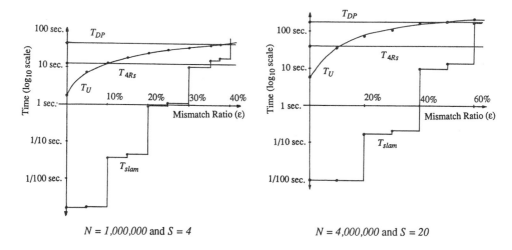

Figure 4. Timing Plots for Queries of Length $P = 80$

Observe from Figure 4 that for the case where $S = 4$, our algorithm is three orders of magnitude faster than any of the others when $\varepsilon < 10\%$. It is two orders of magnitude faster when $\varepsilon < 20\%$, and a single factor of 10 faster when $\varepsilon < 30\%$. Moreover, it crosses over with the best algorithms in the 30-40% range of ε exactly as suggested by the curve for $pow(\varepsilon)$ given in Figure 2. In the case where $S = 20$, our algorithm achieves slightly more modest factors of improvement for the intervals 0-20% (3 orders), 20-40% (2-orders), and 40-60% (factor of 4). When ε is above 60% it performs considerably worse than the 4-Russians algorithm. Even so, note that an all-against-all comparison of this 4 million residue database would take 6 minutes for 0-20% thresholds, 2.6 hours for 20-40% thresholds, and 23 days for 40-60% thresholds. The databases for the timings above were randomly generated and so one would naturally ask if the algorithm is still effective on the data arising in practice. In additional experiments over a 3 million residue of the PIR protein database [28] we found that despite significant skew in the occurrences of $(\log_S N)$-tuples, the algorithms speed was only about 10% slower than in the random case.

Generalizing to Weighted, Local-Alignment Searches

Recall that generalizing from approximate keyword searches to weighted, local-alignment searches consists of (1) permitting arbitrary scores for substitutions and gaps, and (2) considering all substrings of the query (of sufficient length). The first relaxation to generalized Levenshtein measure does not require any significant conceptual changes to our approach. The definition of neighborhood is unchanged and the division Lemma still applies. The only difficulty is that the algorithm for generating a neighborhood must accommodate the fact that edit operation costs can be negative. For V in the D-neighborhood of W, this implies that it is now possible that for some prefix of V, all prefixes of W align to it with a score greater than D. In other words, the monotonicity properties of the underlying dynamic programming matrix no longer hold when negative weights are permitted. However, monotone bounds on each matrix entry can be effectively created by precomputing a lower bound on the best completion of an alignment to a prefix of W. These monotone bounds can then be used in the neighborhood generating algorithm alluded to earlier in the paper.

While the algorithm extends to weighted scores quite easily, we can no longer make formal statements about the running time of the algorithm. The arguments about the size of neighborhoods specifically exploited the properties of simple Levenshtein difference. It may be possible to prove something formally about neighborhood sizes in terms of the mean and variance of substitution scores. This is an area that may yield fruitful results.

The greatest difficulty comes in trying to generalize to substrings of the query. This is because the division lemma no longer applies. Without knowing a priori where the local alignment will begin and end within the query string, organizing a parsimonious decomposition of the query for which the division lemma can be applied seems difficult. It is the case that a division in which the query is broken into every substring of length $2^k \log_S N$ for some k, does the job but adds a factor of $O(\log^2 N)$ to the algorithm (when unit costs are assumed). While theoretically this doesn't appear to be much, in practice it translates to a factor of 100 or more. We continue to explore this issue.

Table 1. Preliminary Query Times in Seconds (N=3 million, P=100)

R (match ratio)	0	0.5	1.0	1.5	2.0
Periodic Clock Protein	225	125	42.5	18.5	6.0
Cytochrome C	245	80	27.5	12.0	3.5
Hemoglobin α-Chain	100	55	20	9.5	2.5

In the interim, the following ad-hoc algorithm was implemented in order to present some preliminary empirical results. The approximate keyword matching algorithm was modified to produce neighborhoods with respect to the Dayhoff-matrix scores at 120 PAMs [1]. This scheme is the one frequently used by BLASTA and FASTA searches. Moreover, as is customary for such schemes, the software was also modified to maximize score, rather than minimize it. With this shift the problem now becomes one of finding matches whose score-to-length ratio is *greater* than a given *match ratio, R*. The software was run against an older version of the PIR protein database [28] that had about 3 million residues of data. For N of this size, $T = \log_{20} N = 5$. As yet we do not have software that implements a complete search strategy but we have timed the following simple exploration. Given a threshold R, we search for all $(R+2)$-matches to a T-tuple of the query, see if these can be extended to $(R+1)$-matches to a $2T$-tuple, and finally if these can be extended to R-matches to a $4T$-tuple. Table 1 below shows our results on the SparcStation 2 described earlier. In each case we searched against only the first 100 residues of each sequence in order to not skew times by the query length. These times are at least encouraging. Experience with a

Dayhoff-matrix scoring scheme reveals that it is not nearly as sharply discriminating as the simple unit cost scoring scheme, i.e., many more matches will occur by chance at an "interesting" level of R. Nonetheless, if a good extension strategy can be worked out, it is clear that the timings below portend a significant increase in speed over the Smith & Waterman algorithm.

REFERENCES

1 Dayhoff, M.O., W.C. Barker, and L.T. Hunt. 1983. Establishing homologies in protein sequences. Methods in Enzymology 91:524.

2 Smith, T.F., and M.S. Waterman. 1981. Identification of common molecular structures. J. Mol. Biol. 147:195.

3 Lipman, D.J. and W.R. Pearson. 1985. Rapid and sensitive protein similarity searches. Science 227:1435.

4 Altschul, S., W. Gish, W. Miller, E. Myers, and D. Lipman. 1990. A basic local alignment search tool. J. of Molecular Biology 215:403.

5 Roberts, L. 1989. New chip may speed genome analysis. Science 244:655.

6 Chow E., T. Hunkapiller, J.C. Peterson, M.S. Waterman, and B.A. Zimmerman. 1991. A systolic array processor for biological information signal processing. Proc. International Conf. on Supercomputing (ICS-91).

7 Levenshtein, V.I. 1966. Binary codes of correcting deletions, insertions and reversals. Soviet Phys. Dokl. 10:707.

8 Sellers, P.H. 1980. The theory and computation of evolutionary distances: pattern recognition. J. of Algorithms 1:359.

9 Ukkonen, E. 1985. Finding approximate patterns in strings. J. of Algorithms 6:132.

10 Mount, D. and E.W. Myers. 1986. Computer program for the IBM personal computer that searches for approximate matches to short oligonucleotide sequences in long target DNA sequences. Nucl. Acids Res. 14:1025.

11 Landau, G.M., and U. Vishkin. 1986. Introducing efficient parallelism into approximate string matching and a new serial algorithm. Proc. 18th Symp. Theory of Computing 220-230.

12 Myers, E.W. 1986. Incremental alignment algorithms and their applications. Technical Report TR86-22, Dept. of Computer Science, U. of Arizona, Tucson, AZ 85721.

13 Galil, Z., and K. Park. 1990. An improved algorithm for approximate string matching. SIAM J. on Computing 19:989.

14 Ukkonen, E., and D. Wood. 1990. Approximate string matching with suffix automata. Technical Report A-1990-4, Dept. of Computer Science, U. of Helsinki.

15 Wu, S., U. Manber, and E.W. Myers. 1991. A subquadratic algorithm for approximate limited expression matching. Technical Report TR92-36, Dept. of Computer Science, U. of Arizona, Tucson, AZ 85721 (submitted to Algorithmica).

16 Ukkonen, E. 1992. Approximate string-matching with q-grams and maximal matches. Theor. Comp. Sci. 92:191.

17 Chang, W.I., and E.L. Lawler. 1990. Approximate matching in sublinear expected time. Proc. 31st IEEE Symp. Foundation of Computer Science 116-124.

18 Ehrenfeucht, A., and D. Haussler. 1988. A new distance metric on strings computable in linear time. Discrete Appl. Math. 20:191.

19 Wu S., and U. Manber. 1992. Fast text searching allowing errors. Comm. of the ACM 35:83.

20 Boyer, R., and J. Moore. 1977. A fast string searching algorithm. Comm. of the ACM 20:262.

21 Gonnet, G.H., M.A. Cohen, and S.A. Benner. 1992. Exhaustive matching of the entire protein sequence database. Science 256:1443.

22 Weiner, P. 1973. Linear pattern matching algorithms. Proc. 4th IEEE Symp. Switching and Automata Theory 1-11.

23 McCreight, E.M. 1976. A space economical suffix tree construction algorithm. J. of the ACM 23:262.

24 Blumer, A., J. Blumer, D. Haussler, A. Ehrenfeucht, M.T. Chen, and J.Seiferas. 1985. The smallest automaton recognizing the subwords of a text. Theor. Comp. Sci. 40:31.

25 Myers, E.W. 1990. A sublinear algorithm for approximate keyword searching. Technical Report TR90-21, Dept. of Computer Science, U. of Arizona, Tucson, AZ 85721 (to appear in Algorithmica).

26 Manber, U., and E.W. Myers. 1990. Suffix arrays: A new method for on-line string searches. *Proceedings of the First ACM-SIAM Symposium on Discrete Algorithms* 319-327. Also to appear in SIAM J. on Computing.

27 Myers, E.W. 1986. An O(ND) difference algorithm and its variants. Algorithmica 1:251.

28 George, D.G., W.C. Barker, and L.T. Hunt. 1986. The protein identification resource. Nucl. Acids Res. 14:11.

PATTERN RECOGNITION IN GENOMIC AND PROTEIN SEQUENCES: A SURVEY OF STATISTICAL VALIDATION PROBLEMS

Jens G. Reich

Max-Delbrück-Centrum für Molekulare Medizin
Berlin-Buch, Germany

SCOPE OF THIS ARTICLE

This text is addressed to biologists working with DNA or protein data bases, but without enthusiasm for mathematical detail. It is intended to expound the principles of significance theory of homology studies and pattern search without the details of formalism. Complete avoidance is impossible: One cannot explain mathematics without any mathematics.

WHY STATISTICS?

Every year letter agglomerations of genomic texts are produced world-wide in the range of dozens of millions. A substantial fraction of this becomes published and submitted to the data banks. The rate of production of genomic information has by far surpassed any capacity to establish its "sense" by the traditional method of experimental study. Only a tiny fraction of genomic facts can be studied in great detail, only a small subset of proteins can be synthesized, crystallized, studied with X-rays, tested in cellular circumstances etc. This fraction has to supply models, and much of the additional knowledge has to be gained by heuristic comparison, at least in the initial attempt to separate the wheat from the chaff.

The majority of genomic and derived information has to be screened by computer analysis. A wealth of software is available for prediction of which information might be important and need further experimental study, while other information might less likely contain signals or might even be senseless at all.

Thus we face screening through data collections of a size that will soon surpass the gigabyte range each time when we want to look for subsequences similar to a given new query sequence. The goal is to predict, with some degree of confidence, the possible function or structure of the query sequence, to which family it may belong etc. This is where statistics comes into play, but in a different spirit as usual. Mostly statistics is called to establish the characteristics of mass events, for instance by mean and standard deviation of a measured quantity, and to construct tests on this information. In sequence comparison the intention is different. We wish to exclude the mass phenomena. We are not interested in all the weak "homologies" that occur by coincidence. Very rare events are to be studied, and what we need is a measure of how surprising a coincidence is. We will see that this requires a statistical reasoning other than usually applied. The screening through large data bases will be envisaged as a succession of random events, although in fact nothing is really random there. We study the fluctuation of certain parameters (like percentage of identity, or like score values seen in a gliding window), and establish limits beyond which any coincidence (high similarity or score or whatever), although still possible, is so unlikely, so surprising that it may give valuable cues for further experiments.

A DATA BASE PARADOX

At the start, there is neither statistics nor randomness. Assume you compare two sequences, letter by letter, each consisting of 100 amino acids. By pure chance you will encounter identical letters in identical positions. Identity of 5% up to 15% is not surprising, whereas a percentage of 80% is strongly indicative of close kinship. In between, perhaps between 20 and 40%, we have a certain region that has been aptly named a "twilight zone" [1, 2, 3, 4].

The twilightiness of a finding depends on certain external circumstances. Most important is the length of the compared sequences. It is obvious that one identical amino acid in a tripeptide alignment is less surprising than 20 identical ones among a series of 60. The binomial distribution teaches us that the former event has a probability of about 14% while the latter is in the range of 10^{-10} percent when the amino acids are approximately equifrequent. The average "expectation" of identical scores is about 5%, and the envisaged peak of 33% becomes the more unlikely the longer the series of "tosses" is.

However, we soon encounter situations where our intuition is not as certain as in this elementary case. For instance, assume a characteristic pattern of 8 amino acids known as active center of an enzyme. The sequences from different species and isoenzymes show, say, between 5 and 8 identical amino acids (62 to 100%). We wish to recognize the pattern, if present, in a new query sequence. The average chance to run into a specified amino acid in one position is 0.058 (sum of squared frequencies of all 20 amino acids). Therefore we expect around 8 times 0.058, i.e. less than 1 identity in a octapeptide sequence that has nothing to do with the pattern. By chance there may occur a few more identities in the 8 places, but more than 4 has a vanishing probability of $1.6 \cdot 10^{-4}$.

Now imagine we have sequenced an open reading frame of 400 amino acids and suspect our pattern somewhere in this sequence. The log(n)-law that we are going to expound in detail predicts with a probability of 6% at least one octapeptide with 5 or more out of 8 to be identical, which is considerable. Searching a data base of 1 million amino acids, we are almost sure to find one or more patterns of 62% or more identity, by pure coincidence, even if our active center is not present. Even 6 or 7 identical amino acids have a considerable probability in such a data heap, only 8 out of 8 is still unlikely (with p=0.058, $p^8 = 1.3 \cdot 10^{-10}$).

The paradox is obvious: Why on earth should it make a difference whether I fished my sequence out of a million instead of out of a few candidates? Active center is active center, and this does not depend on the background noise, or?

There is no difference, of course, provided you are sure that 6 or 7 identical amino acids out of 8 determine structure and function of the oligopeptide in any environment such that it cannot help but becoming an active center. But just this is often in serious doubt. Kabsch and Sander [5] have documented that the same pentapeptide may occur in completely different environment, namely β-strand or α-helix. In many proteins amino acids are interchangeable without harm to the function, so large data bases may contain the same thing in slightly different clothes as well as different things that look very much alike.

The statistical test is not set up to prove or refute any hypothesis. Its main function is **to gauge the degree of surprise of a coincidence**. A further function is to help the selection of screening parameters in computer search. If they are too strict, nothing may turn up, and you may resign prematurely. If they are too loose, you may be drowned with false positives among which you do not find out the really surprising facts. You must calibrate your search in order to know where the mass phenomena prevail and where the real surprise sets in. And this certainly depends on the volume of the data collection on which the screening is done.

SCORING MATRICES AND PROFILES

Recording identities within two aligned sequences is only a special case of a more general and widespread method: comparison by similarity scores. In the general case for each letter pair in the alignment a numerical value for that encounter is read off from a scoring matrix, and the sum of these individual values along the alignment is the similarity score for the sequence pair. Identity counting is a special case of a unitary scoring matrix with unity in the diagonal (corresponding to two equal letters) and zero elsewhere (corresponding to non-identical letters).

Scoring matrices for sequence comparison have usually been derived from two different, but related principles:
- physicochemical similarity of the aligned residues
- evolutionary closeness or distance of the aligned residues

Some authors have used exclusively physicochemical principles, such as the propensity of an amino acid to be found in certain structure elements of proteins [6-9]. Others have derived exclusively from evolutionary reasoning, for example by counting the distance between two amino acids by the number of nucleotide replacements necessary for the transition from one codon to the other one. Dayhoff's well-known PAM matrices [10] stem from a mixed application of both principles, because selection of a position during evolution depends on both processes, mutability as well as function in the cell. Also the method of profile analysis in its original form Gribskov et al.[11] combines both principles, as the scores were calculated as products of weights related to PAM similarity and frequency in a specified profile.

There is a wealth of scoring schemes in the literature [12-17]. For our survey we are interested in the fact that all of them are derived from an implicit statistical model. In some cases the mere frequency of a letter decides the score, or its logarithm. A sequence of letters has a score proportional to the sum of the logarithms of the individual residues, in accordance with the rule that the probability of a sequence is calculated as product of the probabilities of the individual (statistically independent) events.

More frequent are scores based on the comparison of two probabilities of the same letter in different circumstances. If a letter occurs with probability p in a specified sequence family, and with probability q in a random sequence (or, for that matter, in a data base of unrelated sequences) then $\log (p/q)$ is a natural score for its information value, the so-called log odds or log likelihood value. It expresses the ratio of probabilities for two different situations. The score becomes positive and the higher, the more a letter is characteristic for the specified set; if it occurs less frequently than in the background set, then the score becomes increasingly negative. One sees that there is usually an implied probabilistic model behind the usage of scores, hence the further pursuit of the statistical concept has nothing artificial in it. A minor problem with log-likelihood is that care has to be taken of the case that a letter does not appear in one or both sets. The logarithms go to infinity, and scores become indefinite or infinite - both not in accordance with what we expect them to show. The problem is usually circumvented by assigning, instead of zero, a small frequency greater than zero (practically zero) as likelihood of an event.

The extension of the information principle to probabilities involving more than one letter is straightforward. A letter pair in an alignment occurs with frequency f_{ij} in the pattern and with frequency q_i and q_j in the random set, then $\log (f_{ij}/q_i/q_j)$ is the adequate score. Altschul [18] has published detailed arguments for the usage of information measures in sequence comparison.

Whatever the derivation of a scoring scheme, in the mathematical treatment all scores have the common property that they may be treated as a random variable that may attain certain values (the scores) with a defined probability. Sums of individual scores are therefore amenable to the treatment of sums of random variables under a specified statistical model. McLachlan [19] has introduced this principle of statistical treatment into the methodolgy of sequence comparison.

Profiles [11] are scoring schemes depending specifically on the position of a letter in a pattern. This is a typical requirement for pattern recognition, while in evolutionary studies of whole sequences it was acceptable (or inevitable) to use the same matrix (PAM matrix or whatever) for every position of a sequence.

GAUGING SURPRISE AS AGGREGATE SCORE: FIXED WINDOW VS. FREE SEGMENT

Scores for the appearance of letters in a sequence as well as for letter pairs in an alignment are collected as sum over all positions. The further the score moves away from the mean of the mass phenomena (from the "random" events), the more surprising is the result.

There are two ways to count an aggregate score defined over a sequence. One is with a fixed window of defined width, which glides over the alignment or sequence. In each possible position the score is recorded. This method is most suitable when a sequence pattern of defined length is looked for, or a binding site or otherwise a specified motif.

A second counting method does not work with windows of defined length. One is just looking for "something surprising", i.e. for a high-scoring segment somewhere in the sequence (or dot matrix) without prescribing its length or position. This method suggests itself when one is looking for a partial or complete homology between sequences, without knowing in advance where and how long.

These two methods are related, but there are differences which bear on the mathematical treatment of the possible significance of a high score. Consider some examples. In the following pattern

1 1 1 0 1 1 1 1 1 1 1 1

a fixed window of width 12 will count a score of 11. The free segment method will come out with 11 (one mismatch), and alternatively with two segments of score 8 and 3 (without mismatches).

In the following situation:

0 0 0 0 1 1 1 1 0 1 1 1 1 1 0 0 0 0 0 0

a fixed window of width 20 will turn out a score of 9 (not very high, perhaps) while the free segment method will show a possibly surprising 9 (one mismatch) or two uninterrupted segments of score 4 and 5. In the following case, which is a permutation of the previous one:

1 0 0 0 1 1 0 1 0 0 1 0 1 1 0 1 0 0 0 1

the fixed 20-window exhibits the same score of 9, while the free segment comes out with insignificant chaff of a few points only.

The method of free segments has the advantage of being more flexible. Also its extreme value statistics is easier to derive (as we will see). The disadvantage is that outcomes of different segment length tend to be clustered (as in the first example the 11 score with 1 mismatch uses the same section as the 8 score without mismatches) and are therefore not statistically independent.

This leads us to a general rule of sequence comparison: Neighbouring aggregate scores within one diagonal of the incidence matrix tend to use the same elemental scores and are therefore positively correlated, while aggregate scores from neighbouring diagonals are practically independent. For illustration, in the following incidence matrix:

	A	T	T	T	C	C	A	G
A	1	0	0	0	0	0	1	0
T	0	1	1	1	0	0	0	0
C	0	0	0	0	1	1	0	0
T	0	1	1	1	0	0	0	0
C	0	0	0	0	1	1	0	0
C	0	0	0	0	1	1	0	0
A	1	0	0	0	0	0	1	0

in the first diagonal (elements underlined) there are several length-4-windows with a score of 3 or 4, while in the immediately neighbouring diagonal (elements in bold face) only scores of 1 or 2 appear. Thus, gliding along a diagonal produces a positively correlated score series, neighbouring diagonals do not (for details, see [20]).

ORIENTATION: THE LOG(N)-LAW

If a certain scoring scheme is applied repeatedly to the subsequences in a data base there will be a tendency that the best score increases slowly with the number of tests. Thus, the level beyond which a score becomes significant will grow slowly. Let us study a simple example in order to understand the nature of this phenomenon.

Consider a search for a short pattern in a very long data base. Fig. 1 shows that this is equivalent to a search for x dots in one of the N diagonals of a dot matrix produced by gliding with a x-letter-window over the database and summing up the dots seen. Assume an average dot density of p in the whole dot matrix, so we estimate the probability to meet a full diagonal (a pattern hit) by

$$\text{Probability (x dots in a given diagonal)} = p^x,$$

which may also be written in the notation

$$\text{Prob(x dots)} = \exp(-x\lambda) \tag{1}$$

with

$$\lambda = \ln\left(\frac{1}{p}\right)$$

being a characteristic number expressing the change of base of logarithms from p to e.

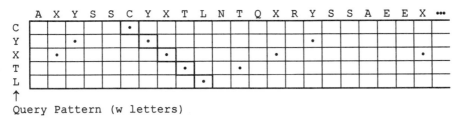

Figure 1. Data base, query pattern, resulting dot matrix and scores. The score of the search in a given position is the sum of dots in the corresponding diagonal. See the identical subsequence beginning at pos. 6 of the data base: its score is 5 dots. The matrix allows also to calculate the dot density (probability of identical letters) as ratio of dot-containing over total entries in the whole dot matrix.

As a consequence, the probability to obtain a diagonal with less than x dots (a non-hit) is

Probability (less than x dots in a diagonal) $= 1 - p^x$

Now we test not one diagonal, but a long series of N diagonals and may assume, as mentioned in the previous section, that the contents of neighbouring diagonals are statistically independent of each other. The probability Q of seeing none of the N diagonals filled with a hit of x dots ist therefore the N-th power of the individual failure probability:

$$Q = \text{Probability (no diagonal with a hit)} = \left(1 - p^x\right)^N \tag{2}$$

and may be approximated by

$$Q \approx \exp\left(-N \cdot p^x\right) \tag{3}$$

Hence the probability *(1-Q)* of at least one full dot diagonal grows according to an exponential law. Its characteristic value, where *1-Q* equals *(e-1)/e = 0.63*, is reached when $N \cdot p^x$ equals unity. In a database whose volume N surpasses the reciprocal of the hit probability we may predict with increasing certainty that random pattern hits will be present.

On the other hand, the longest series of dots in a database of N window positions may be calculated. Prescribe the desirable statistical confidence (say: *1 - Q* be 0.99) and solve for x :

$$x = \frac{\ln N - \ln \ln \dfrac{1}{Q}}{\ln \dfrac{1}{p}}$$

which is to say that with confidence 99% we may expect a run of at least x dots when screening N window positions. The length of such a run grows with the logarithm of N (to base *1/p*). Note that x as just calculated is not an integer value, so what we may expect as dot run is the integer closest to this calculated value.

It turns out that the *log-N law* is a rather general relationship: High scores tend to grow with the logarithm of the data base visited.

Consider a numerical example. If Q is set to 0.99, as explained, and p = 0.058 (frequency of amino acid coincidence when searching a data base of protein sequences), and N = 8 millions (number of residues at present in those databases), we may calculate

$$x = \frac{\ln\left(8 \cdot 10^6\right) - \ln \ln(100)}{\ln \dfrac{1}{p}} = \frac{15.9 - 1.53}{2.85} = 5.05$$

In other words, the longest uninterrupted dot series obtainable by chance is a pentapeptide, or possibly (but with rapidly decreasing probability) somewhat longer, a hexa- or heptapeptide.

STANDARD FORM OF THE EXTREMAL VALUE DISTRIBUTION: ITS UNIFORM TAIL

Relationship (3) may be written in a standardized form. We introduce the form (1) for p^x. With this we get, from (3),

$$Q = \exp\left\{-N \exp(-x\lambda)\right\} \tag{4a}$$
$$= \exp\left\{- \exp(\ln N - x\lambda)\right\} \tag{4b}$$

Instead of x (the number of dots) we introduce u, the number of uninterrupted dots exceeding the characteristic value of $ln\ N$, measured in units of λ

$$u = x\lambda - \ln N$$

and obtain

$$Q = \exp\{-\exp(-u)\} \qquad (5)$$

which is the standard form of the extremal value distribution Gumbel [21], see treatment by [22, 23]. It can be shown that it has an expected mean of $\gamma = 0.577$ (Eulers constant), a standard deviation of $\pi/\sqrt{6}=1.28$ [24], and that its 99%-quantile lies at 4.6. The most important aspect of this is that only the expected mean of x, not the scatter, grows with $logN$. With increasing N the distribution of extreme scores is shifted towards the right, but without distortion of the shape: The tail of the distribution is unchanged. The prediction of highest scores remains precise in spite of the increasing database volume.

SEARCH FOR THE LONGEST SERIES OF HITS, WITH INTERRUPTION

The basic idea of the $log(N)$-law, as outlined in the previous section, was elaborated in much detail and applied to pattern recognition and homogenity search in sequences by two groups of mathematicians, the senior authors being S. Karlin and M.S. Waterman. The results are summarized in several recent surveys [25, 26]. We mention here only the main results.

Erdös and Rényi [27, 28] proved an exact version of the $log(N)$ law for series of independent coin tosses with p between zero and unity. Arratia et al. [29] show that in matching two sequences the longest match grows with $log(nm)$, i.e. of the product of the lengths of both sequences, see also [30-33]. Karlin et al.[26] have a formula for the occurrence of long repeats in sequences. Arratia et al. [34] give an expression for longest matches allowing for a few allowed mismatches in between. There are also solutions for non-independent letter succession (Markov model, [35]), for multiple hypothesis testing [36, 37] and for repeated occurrences of the same rare event or for clusters of different rare events [26].

All this formalism is applicable using hand-held calculators. The results hold strictly only for very long sequences, but the approximation is good enough for sequence comparison as encountered in daily practice. They do not cover strict fixed-window search, and are not directly applicable to the possibility of indels (insertions or deletions) in the alignment. The formalism avoids application of the normal distribution in the statistical treatment (z-scores as indicator of significance etc.), which would not be correct when dealing with extremely rare events, as they occur with pattern search in large data bases (see below).

SCORING SCHEMES: STATISTICS OF EXTREME SCORES

So far only the special case of counting identities has been covered. The more universal situation, with a full scoring matrix, was shown by Karlin and co-workers [38-41] to be treatable by the same general formalism.

The basic model of this theory is a set of possible individual scores, s_i, each of which may occur with a given probability p_i, so that a very long series of N realizations of this scoring scheme results. The question is in the same spirit as previously: What is the maximal aggregate score value of any segment in such a series?

A simple reflection shows that certain restrictions of the numerical values of the score matrix are necessary to answer this question in a meaningful way. If there are only non-negative scores, s_i, it is clear that the whole chain is the segment with highest score, because it can only increase. The scoring scheme must instead produce a "negative drift", which means that it needs negative scores as well as positive ones (at least one of each), and that the expected average score, the sum over all $p_i \cdot s_i$ be negative.

This requirement may in many cases be met by a simple transformation. Our standard example, counting identities as unity and non-identities as zero, qualifies for this theory, if we replace it by the scores +1 for identity and -1 for non-identity, when *p(identity)* < *0.5* (as in most applications). This produces score sums between +x and -x instead of +x and 0, but leaves the relative order of scores in a given search unchanged. If *p(identity)* > *0.5* then the scores must be shifted further to the left (say +1 for identity and -2 for non-identity) until a negative drift entails, according to the above requirement. Any scoring scheme with a non-negative drift can be transformed to meet the requirement by simply subtracting a constant greater than this expected value from the scores.

For scoring schemes fulfilling this requirement Karlin et al. [38] have proven that the maximal segment score *M(n)* in a series of n independent evaluations is in the order of:

$$M_{(n)} = \frac{\ln(n)}{\lambda^*}$$

which is the same *log(n)-law* that we have considered already for series of dot runs. The only complication is that this time there is no explicit expression for the characteristic modulus λ^*. Instead, it is obtained as unique positive solution of the equation

$$\sum_i p_i \exp\{\lambda s_i\} = 1$$

The factor λ^* is applied to all scores such that the expected value of $\exp\{s_i\}$ becomes unity. It lies in the interval between zero and $\ln(1/p_{max}) / s_{max}$, with s_{max} being the maximal score of the matrix scheme, p_{max} being the pertinent probability. For instance, for $p_{max} = 0.1$ and $s_{max} = + 8$, the modulus is between 0 and 0.288 and may be easily calculated by a simple non-linear equation solver.

Karlin & Altschul [38] give not only the log(n)-law for the maximal score, but also its distribution, which again is an extreme value distribution:

$$\Pr obability(M_{(n)} - \frac{\ln(n)}{\lambda^*}\langle x\rangle = \exp\{-K^* \cdot \exp(-\lambda^* x)\} = \exp\{-\exp(-u)\}$$

with

$$u = \lambda^* x - \ln K^*$$

The quantity K^* is calculable as a converging series. Note the complete analogy of these expressions to formulae (4a) and (5). For a similar treatment, see [42-44].

OPTIMAL SCORING SCHEMES

Karlin & Altschul [38] and in more detail Altschul [18] have shown that the theory of maximal segment score may define an optimal scoring scheme. Their basic argument is that any specified scoring scheme tends to select maximal segment scores of a distinct letter composition, which they call target distribution.

A simple example may explain this. Consider a sequence consisting of two elements, 3-H-bond nucleotides A or T (letter S) and 2-H-bond nucleotides (letter W), occurring with probabilities $p_S = 0.4$ and $p_W = 0.6$. If we assign scores $s_S = +1$ and $s_W = -1$ then it is clear that a high scoring segment will tend to contain more than the random 40% S letters and less than

60% W letters. [38] prove a theorem from which the target distribution of our example may be calculated to be

$$q_s = p_s \exp\{\lambda^* \cdot s_s\} \text{ and } q_w = p_w \exp\{\lambda^* \cdot s_w\}$$

which may be directly calculated to be (as $\lambda^* = \ln (p_w/p_s)$ in this case): $q_s = 0.6$ and $q_w = 0.4$. The maximal segment score tends to be $\ln(n)/\lambda^*$ and its composition is 60% S and 40% W, just a mirror image of the average composition.

This argument may be turned around for the purpose of pattern recognition. If p_i is the set of probabilities that characterize the letter composition of a "random" sequence and q_i the set characterizing the letter composition of the target pattern, then the optimal scoring set to find just this class of patterns is given by

$$s_i = \frac{\ln\left(\frac{q_i}{p_i}\right)}{\lambda^*}$$

This principle has been applied (implicitly) by [10] in deriving their PAM mutability matrices.

OPTIMAL SCORES IN PROFILE ANALYSIS

When definite motifs or patterns are to be identified by a scoring scheme (cf. [7]), we need a vector of scores for each letter of the alphabet. However, these scoring vectors are position-dependent; they vary from position to position in the pattern. We may still apply optimal scores in the sense as previously outlined, i.e. we take

$$s_i \text{ proportional to } \ln\left(\frac{q_i}{p_i}\right),$$

where q_i is the frequency of the i-th letter in the pattern position and p_i the frequency in a random sequence. Letters that do not occur at all in a position must be treated as very rare letters, i.e. assigned a suitably small probability > 0. Such a scoring scheme will select high-scoring subsequences that tend to be similar to the target pattern (or consensus sequence) of the motif which is being looked for.

SCORING THROUGH A GLIDING WINDOW

Assume that we have a pattern of n positions. For each position a scoring vector be given, assigning a specific score to the occurrence of each letter. To begin with a simpler case, assume integer-valued scores (zero and negative integers also permitted). Be there a set of probabilities describing the behavior of the "random" sequence (as model of the average data base subsequence). In each position of the window gliding over the data base we may record a score, and the problem is how often high scores occur by fortuitous coincidence. Only scores that are extremely unlikely are sufficiently surprising as to justify the hypothesis that this is a true occurrence of the pattern.

What we need at first is the probability distribution of the possible values of the aggregation scores in one random subsequence. It may be calculated by the method of probability-generating polynomials, as applied to sequence analysis by [19, 45]. Table 1 gives a simple illustration of this principle, which is easier to understand than the mathematical formalism. It is not difficult to write a program that performs such repeated polynomial multiplication, although considerable computer memory may be required (e.g. for integer scores between -10 and +10 and a pattern of 30 positions, we get probabilities for aggregate scores between -300 and +300). One must be cautious of arithmetic underflow or overflow.

What we get in this way is a vector of probability values for each possible aggregate score from minimum to maximum (say -300 to +300). We may now calculate the probability of any rare event by summing up the coefficients for all scores greater than x. This is the tail probability of the distribution, also called the quantile.

Having obtained this small tail probability α for a rare (surprising) event in one window, we may expand to a large number N of windows and arrive at an analogon of the *log(N)-law* along the same argument as used above for the geometric distribution. The probability of finding at least one "hit" (rare event from the tail of the distribution) in a large series of N window inspections is given by:

$$\text{Prob (seeing at least one hit)} = 1 - (1 - \alpha)^N,$$
$$\approx 1 - \exp(-\alpha \cdot N).$$

As a definite score value corresponds to each α , this relationship may be used to construct the distribution of scores under a large series of inspections of a data base through a gliding window.

Table 1. Method of probability-generating polynomials (illustrated)

score	-1	0	+1	polynomial
pos. 1	0.2	0.5	0.3	$(0.2\ x^{-1} + 0.5\ x^0 + 0.3\ x^1)$
2	0.8	0.05	0.15	$(0.8\ x^{-1} + 0.05\ x^0 + 0.15\ x^1)$

$$\Pi = 0.16\ x^{-2} + 0.41\ x^{-1} + 0.295\ x^0 + 0.09\ x^1 + 0.045\ x^2$$

Example: Prob (sum score = -1) = 0.41

Given a "pattern" of two positions. In each of them the scores -1, 0 and +1 are possible, with different probability (e.g. in pos. score -1 with probability 0.2, etc.). This may be expressed symbolically as polynomial, where the coefficient expresses the probability and the power of x the score (e.g. 0.2 x^{-1} in the example just quoted). The probability to obtain sum scores from both positions is calculated by multiplying both polynomials, resulting in Π. Again, the coefficient of an aggregate score is given by the coefficient at the corresponding power of x (e.g. 0.41 x^{-1}).

APPROXIMATION FORMULA: BEWARE OF NORMAL DISTRIBUTION!

The tedious calculations necessary in all but the most elementary cases has prompted the application of simpler estimation schemes. The most popular one of this, introduced by [10], is the formalism of z-scores.

It is simple. A given scoring scheme is applied either to a statistical model, or to a Monte Carlo simulation (e.g. scrambled sequences), or to the bulk of the data base sequences, and the sample mean and standard deviation is calculated. Each query score may then be expressed in z-units. A z-value says how many standard deviations the obtained value is above the mean. After a lookup into a table of normal distribution this z is then transformed into an error probability (= probability of false positives). The more standard deviation units the finding is away from the mean the less probable is it that this is a chance event.

There is a principal flaw in this method that comes to light just in the case of vary rare (very large) deviations. The fact is that the far tail of the normal distribution goes dramatically too quickly to zero. This tail may be (large z) approximated [46] by:

$$\frac{\exp\left\{-\dfrac{z^2}{2}\right\}}{\sqrt{(2\pi)} * z}$$

It is seen that this expression approaches zero much more rapidly than e^{-z} which is the usual exponential decay of aggregate scores, binomial tails, geometric series and similar objects. Thus for large z the normal approximation will severely underestimate the probability of large excursions. Only for smaller deviations ($z<1$) it can be shown that the approximation by the normal distribution is correct [48].

As example let us consider the binomial distribution with a comparison of two sequences each 100 amino acids long. The dot density is 0.058. We ask for the probability to get a score of 20% identity or better. The answer is given by the binomial "tail":

$$\text{Prob } (k \geq 20) = \frac{100!}{k!(100-k)!} \cdot p^k \cdot q^{100-k} = 1.05 \cdot 10^{-6}$$

On the other hand, mean ($= 100 \cdot p = 5.8$) and standard deviation ($\sqrt{\{100 \cdot p \cdot q\}} = 2.33$) say that a value of 20% is 6.1 z-units away from the mean. Its tail probability under normal distribution is about $6 \cdot 10^{-10}$, i.e. by more than 3 orders of magnitude smaller than the true value. A screening in a 10-million-database is predicted by the binomial distribution to turn out an average of 10 random hits (10 millions times $1.05 \cdot 10^{-6}$), and the probability to see no hit is $3 \cdot 10^{-5}$. The normal approximation makes a quite different and fundamentally wrong prediction. It says that the probability to see no hit is more than 99% even with 10 millions because the tail probability is so vanishingly small.

Thus the correct value predicts overflooding with false hits, while the approximation will declare all of them as significant, indeed a misleading statement. The normal approximation is invalid for very large deviations from the mean! And "very large" begins at $z=3$ or $z=4$. Many authors are aware of this pitfall (e.g. [19, 25]), but the method of evaluating z-scores with the normal distribution is still popular. However, it is appropriate to fit an exponential distribution (rather than a Gaussian one) to the data. See [22, 23, 30, 31, 32, 49], for further details.

DIAGONAL SCORING MATRIX WITH FIXED WINDOW WIDTH

If a certain sequence is to be detected in a large data bank then the treatment with maximal segment length (without prescribed window width) is not practical, because the interest is in the presence of a pattern of definite length. The screening is done with a fixed window that has the width of the pattern length. See for example [50, 19, 51].

If we score two classes of letters (those favorable, or compatible, with the pattern and those not) the treatment leads, in essence, to the tail of the binomial distribution, with the tossing repeated a large number of times.

Assume dot density (letter coincidence probability) p , and a window width of W positions. Assign a score of +1 to favorable and 0 to unfavorable letters (the numbers do not matter, any other pair of scores could be taken as well). Consider the probability of a window with high dot density s or more (i.e. $B = W \cdot s$ identical amino acids, where $s > p$). Following equation (11) of Reich and Meiske [20] we obtain

Prob (observed value of dot density in window $\geq$ s)

$$= \exp\{-W \cdot H(s,p)\} \cdot \frac{c}{\sqrt{W}}$$

where

$$H = s \ln\left(\frac{s}{p}\right) + (1-s) \ln\left(\frac{1-s}{1-p}\right) \text{ (entropy term)},$$

and

$$c = \frac{s(1 - p)}{(s - p) \cdot \sqrt{\{2 \cdot s(1 - s)\}}}$$

The value of c is often close to unity (e.g. for p=0.05 and s=0.3). Waterman [25] gives the same formula without $\sqrt{W}$, which overestimates the probability quite considerably (at W = 100 by 1 order of magnitude). In one long diagonal scores are clustered, which leads to a different formula (eq. (9) of [20]).

With this hit probability in one position we may now enter the analysis of repetitive search in a large data base of N window positions. It leads to the double exponential distribution as already encountered (eq.(4)):

(prob of at least one score ≥s in N-database)
$$= 1 - \exp\{-M \exp(-W \cdot H(s, p))\}$$

where

$$M = \frac{N \cdot c}{\sqrt{W}}$$

The characteristic value of the highest score grows, as previously, with log(N).

If the letter probability p changes with every position in the pattern, which is mostly the case, then an approximate treatment with a Poisson distribution is possible, provided that p's are small quantities.

TREATMENT OF GENERAL SCORING SCHEMES WITH FIXED WINDOW WIDTH

When the scoring is independent of the position then this leads to the frequency of large deviations of sums of random variables under a multinomial distribution, which is notoriously intricate to treat analytically. Still more complicated is the realistic assumption, that each set of scores varies from position to position.

These and other complications cannot be treated explicitly any longer. It is recommended to switch to calculating the distribution of scores on a computer. The method of probability generating polynomials, as described by [19] is appropriate. Staden [45] has published algorithms that are convertible into subroutines.

The general idea is unchanged. Calculate the tail probability by multiplying polynomials, and apply the log-N-law of data base search to them.

It is advisable to convert the elementary scores to integer values, The polynomial multiplication would also work with real-valued scores, but the combinatorial explosion of possible aggregate scores (non-integer powers in the "polynomial") would anyway require the collection of nearly equal scores into histogram entries applying to an interval.

THE GAP AND INDEL PROBLEM IN SEQUENCE STATISTICS

The possibility of deletions or insertions is a considerable complication of sequence analysis. It creates two main problems: construction of an alignment becomes difficult, and the statistical significance test is fraught with complications.

The term indel (= deletion in one sequence or insertion into the other when two are compared) indicates that there is a mirror symmetry in the treatment of alignments that do not fit on a letter-by-letter basis onto each other. One sequence's deletion is the same as the other one's insertion, unless there is additional knowledge on the prehistory. The main problem in significance analysis is how to treat such indels. One extreme choice is to treat them like valid letters, allow their insertion at any place of the alignment. Global alignments according to linear programming algorithms [52-58] produce alignments in unrelated sequences where a considerable number of (mostly) isolated gaps is spread over both sequences.

This is not a realistic model. Evolution does not permit gaps or insertions in considerable quantity, and it does not spread them evenly over the sequence. An indel of more than one letter (a loop in one sequence) is much more probable than a series of indels scattered over the sequences. This has been taken into account by allowing gaps in alignments, but assigning a certain penalty for them in the scoring scheme. Widening of an existing gap is less penalized than opening a new one.

The statistical description of this situation is very difficult. Mathematical results are very scarce [59]; some useful results have been obtained by Monte Carlo simulation [60], [49]. Of particular interest is the finding [61] that there is a phase transition: With high gap penalties the log(n)-law holds, while with more liberal low penalties a linear dependence of scores on sequence length obtains.

In pattern analysis the problem may simplify, because the degree of freedom is reduced. A motif is a rather compact thing which allows indels only as rare event and only in certain positions. In other words, there is no combinatoric explosion of the statistical treatment of gaps. Even if deletions or insertions of considerable length come into play, this is possible only in certain places. The motif may in such cases be envisaged as consisting of a few compact submotifs whose distance may vary. Between these submotifs there may occur rather long loops or deletions, the number of indels being no deciding issue. An analogous situation results when exons or other blocks of genetic information are shuffled and/or combined and/or duplicated. Then only a certain succession of these blocks may matter, while the length of the material in between is less important.

Karlin and Altschul [37] solve the statistical treatment of sequences with indels as follows. They say that the simultaneous presence of different submotifs may be treated as collection of independent events. This leads them to a Poisson type distribution of repeated high-scoring events. Afterwards, if there is a certain successive order of submotifs to be heeded, this can be introduced by dividing the probability of rare events by a combinatorial expression that counts only those admitted successions.

The author of this survey has worked with a different method (unpublished), more in the spirit of fixed windows. The production of a certain succession of submotifs is treated as a dice-casting game. A success requires that all submotifs were produced in the correct order. Admissible gap positions imply that the gambler has several tosses to produce the next submotif. The game is lost when this has failed. The probability of getting the submotif is itself given as geometric or related distribution. As the probability of any submotif has to be rather small (in order to be informative at all) one may approximate its chance of occurrence during the game by multiplying with the number of trials admitted. In this way one gets an expression for the probability of the union event of all submotifs under a set of bounded gap lengths. This game is repeated along the database. The expected duration of each game (the majority of games will get lost, the number of rare successes grows as log-N law) may be calculated and permits the transformation from the sequence length into the expected number of games. There is the problem of overlap, but not more serious than in other estimation problems of non-elementary events along the sequence.

Which of the two strategies (restricting loop length or refraining from doing so) is preferable depends on the situation under scrutiny.

ENVOI

Finishing this survey the main points are summarized:

1. Statistical treatment is required for pattern search in large data bases as to refute chance coincidence, to validate high-scoring findings, and to design parameters for a promising search.

2. Pattern searches in data bases that contain 10^6 or more items imply concentration on very rare events and complete neglect of the mass phenomena. When the statistical treatment is to rely on sampling procedures (e.g. unrelated parts of the data base as control, or search sequences scrambled with Monte Carlo) then it must be ensured that the sample is of the same astronomic size as to measure very rare events. This is mostly impractical. With a few thousand random comparisons a confident statistical judgement cannot be obtained for events in the extreme tail (probability 10^{-6} or less).

3. Scores generally grow with the logarithm of the data base volume (*log-N-law*). This means that an exponential type of probability distribution applies in most cases. Very rare events are then anaylized according to a double exponential or extreme value distribution. This is exemplified at the simplest example of a long run of successful coincidences appearing in a very long series of trials.

4. The widely used normal approximation (expressing scores in standard units above the mean) is an invalid approximation under the condition of very large deviations (more than 3-4 standard deviations). The true distribution is close to the exponential one, where the rarest events are more probable than under the Gaussian.

5. Pattern search may be evaluated with the help of scoring matrices. There is a theory suggesting optimal values of scores (the log-odds-method). In evolutionary studies, the same scoring matrix is applied to all positions of the alignment studied, which is a useful simplification, but may break down in pattern search.

In pattern search (predicting binding and active sites, hydrophobic clusters, exon and other motifs, etc.) the score vector for each letter element may change from position to position.

6. In the former case, when one scoring scheme is applied a large number of times, the mathematical theory of longest head-runs in coin tossing or maximal segment length under a scoring scheme is applicable. Karlin and Waterman have given useful explicit expressions for the statistical validation. They rely on the log-N law. The statistical scatter is small and may be described by the double exponential distribution.

7. In the second case of search of a pattern of given length, where a different scoring vector is applied in each position, the method of gliding with a window of fixed width is to be applied. A modified version of the log-N law applies to this case, and again the scatter may be analyzed with a double exponential distribution. Again explicit approximations are possible, with the exception of multinomial scoring schemes. In this case the method of generating polynomials is appropriate.

8. The gap problem is in pattern recognition less formaidable than in global sequence comparison for detecting weak homology. The reason is that gap location and length may be restricted, so that the statistical treatment may include combinatorial corrections to the basic pattern occurrence probabilities.

REFERENCES

1.   Doolittle,R.F., 1986, "Of URFS and ORFS: a primer how to analyze derived amino acid sequences", University Science Books, Mill Valley,CA.
2.   Doolittle,R.F., 1990, Searching through sequence databases, Meth. Enzymol. 183:99-110.
3.   Taylor, W.R., 1988, Pattern matching methods in protein sequence comparison and structure prediction. Protein Engineering 2, 77-88.

4.	Sander,C. and Schneider,R., 1991, Database of homology-derived protein structures and the structural meaning of sequence alignment, Proteins 9:56-68.

5.	Kabsch,W. and Sander,C., 1983, Dictionary of protein secondary structure: pattern recognition of hydrogen-bonded and geometrical features, Biopolymers 22:2577-2637.

6.	Rao,J.K.M., 1987, New scoring matrix for amino acid residue exchanges based on residue characteristic physical parameters, Int.J.Peptide Protein Res. 29:276-281.

7.	Gribskov,M., Lüthy,R., and Eisenberg,D., 1990, Profile analysis, Meth. Enzymol. 183:146-159.

8.	Lüthy,R., McLachlan,A.D., and Eisenberg,D., 1991, Secondary structure based profiles: use of structure-conserving scoring tables in searching protein sequence databases for structural similarities, Proteins 10:229-239.

9.	Bowie,J.U., Lüthy,R., and Eisenberg,D., 1991, A method to identify protein sequences that fold into a known three-dimensional structure, Science 253:164-170.

10.	Dayhoff, M.O.,Schwartz, R.M., and Orcutt, B.C., 1978, A model of evolutionary change in proteins, pp.345-352 in: "Atlas of protein sequence and structure, vol 5, suppl. 3", M.O. Dayhoff, ed., National Biomedical Research Foundation, Washington.

11.	Gribskov,M., McLachlan,A.D.,Eisenberg,D., 1987, Profile analysis: detection of distantly related proteins, Proc.Nat.Acad.Sci.USA 84:4355-4358.

12.	Schneider, T.D., Stormo, G.D. and Gold, L., 1986, Information content of binding sites on nucleotide sequences. J. Mol. Biol. 188, 415-431.

13.	von Heijne, G.1987, Sequence analysis in molecular biology, Academic press, San Diego, CA.

14.	Stormo, G.D., 1988, Computer methods for analyzing sequence recognition of nucleic acids. Ann.Rev.Biophys. Biophys. Chem. 17, 241-263.

15.	Bork,P. and Grunwald,C., 1990, Recognition of different nucleotide-binding sites in primary structures using a property-pattern approach, Eur.J.Biochem. 191:347-358.

16.	States, D.J., Gish,W. and Altschul,S.F., 1991, Improved sensitivity of nucleic acid data base searches using application-specific scoring matrices. Methods (a companion to Methods in Enzymology), 3,66-70.

17.	Gribskov,M. and Devereux,J., 1992, "Sequence analysis primer", Freeman, New York.

18.	Altschul,S.F.,1991, Amino acid substitution matrices from an information theoretic perspective, J.Mol.Biol.219:555-565.

19.	McLachlan,A.D., 1971, Tests for comparing related amino-acid sequences, J.Mol.Biol.61:409-424.

20.	Reich,J.G. and Meiske,W., 1987, A simple statistical significance test of window scores in large dot matrices obtained from protein or nucleic acid sequences, Comp.Appl.Biosci. 3:25-30.

21.	Gumbel,E.J., 1958, "Statistics of extremes". Columbia University Press, New York.

22.	Altschul,S.F., and Erickson,B.W.,1986, A nonlinear measure of subalignment similarity and its significance levels, Bull. Math. Biol. 48:617-632.

23.	Altschul,S.F., and Erickson,B.W.,1988, Significance levels for biological sequence comparison using non-linear similarity functions, Bull. Math. Biol. 50:77-92.

24.	Kendall, M.G. and Stuart, A., 1971, "The advanced theory of statistics, 3d ed., vol.1, p.344", Griffin, London

25.	Waterman,M.S., 1989, "Sequence alignments", in: Mathematical Methods for DNA Sequences, M.S. Waterman, ed., CRC Press, Boca Raton, FL.

26.	Karlin,S., Ost,F., and Blaisdell,B.E., 1989, "Patterns in DNA and amino acids sequences and their statistical significance", in: Mathematical methods for DNA sequences, M.S. Waterman,ed., Boca Raton, FL.

27.	Erdös,P. and Renyi,A., 1970, On a new law of large numbers, Journ. Analyse Math. 22:103-111.

28.	Erdös,P. and Renyi,A., 1975, On the length of the longest head-run, Coll.Mathem.Soc.Janos Bolyai 16:219-227.

29.	Arratia,R. and Waterman,M.S., 1985, An Erdös-Renyi law with shifts, Adv. Math. 55:13-23.

30.	Mott,R.F., Kirkwood,T.B.L. and Curnow, R.N., 1989, A test for the statistical significance of DNA sequence similiarities for application in databank searches. Comp. Applic. Biosci. 5, 123-131.

31.	Mott,R.F., Kirkwood,T.B.L. and Curnow, R.N., 1990, An accurate approximate to the distribution of the length of the longest matching word between two random DNA sequences. Bull. Math.-Biophys. 6,773-784.

32.	Mott,R., 1992, Maximum likelihood estimation of the statistical distribution of Smith-Waterman local sequence similarity scores, Bull. Math.Biol. 54:59-75.

33.	Collins,J.F., Coulson,A.W.F. and Lyell,A.,1988, The significance of protein sequence similarities. Comp.Appl.Biosci.4,67-71.

34.	Arratia,R.,Waterman,M.S., 1989, The Erdös-Renyi strong law for pattern matching with a given proportion of mismatches, Ann.Prob. 17:1152-1169.

35.	Karlin,S. and Dembo,A., 1992, Limit distributions of maximal segmental score among Markov-dependent partial sums, Adv.Appl.Prob. 24,113-140.

36. Waterman,M.S. and Gordon,L., 1990, Multiple hypothesis testing for sequence comparisons, in "Computers and DNA", pp.127-135, Bell,G.I.and Marr,T.G., eds., Addison-Wesley, Reading.

37. Karlin,S. and Altschul,S.F., 1993, Applications and statistics for multiple high-scoring segments in molecular sequences, Proc. Nat. Acad. Sci. USA (submitted 1993, pers. comm.)

38. Karlin,S. and Altschul,S.F., 1990, Methods for assessing the statistical significance of molecular sequence features by using general scoring schemes, Proc.Nat.Acad.Sci.USA 87:2264-2268.

39. Altschul,S.F.,Gish,W.,Miller,W.,Myers,E.W., and Lipman,D.J, 1990, Basic local alignment search tool, J.Mol.Biol. 215:403-410.

40. Karlin,S.,Bucher,P.,Brendel,V,and Altschul,S.F., 1991,Statistical methods and insights for protein and DNA sequences, Annu.Rev.Biophys.Biophys.Chem.20:175-203.

41. Karlin,S. and Brendel,V., 1992, Chance and statistical significance in protein and DNA sequence analysis, Science 257:39-49.

42. Arratia,R.,Gordon,L.,and Waterman,M.S., 1986, An extreme value theory for sequence matching, Ann Stat. 14:971-993.

43. Arratia,R.,Morris,P.,and Waterman,M.S., 1988, Stochastic scrabble: large deviations for sequences with scores, J.Appl.Prob. 25:106-119.

44. Gordon,L., Schilling,M.F., and Waterman,M.S., 1986, An extreme value theory for long head runs, Prob. Th. Rel. Fields 72:279-287.

45. Staden, R., 1989, Methods for calculating the probabilities of finding patterns in sequences, Comp.Appl. Biosci. 3,25-30.

46. Feller,W., 1968, "An introduction to probability theory and its applications, vol.1.", Wiley International, New York-London-Sidney.,

47. Bahadur, R.R., 1960, Some approximations to the binomial distribution function, Ann.Math.Statist. 31,43-54.

48. Bahadur,R.R., 1971, "Some limit theorems in statistics", SIAM, Philadelphia.

49. Smith,T.F.,Waterman,M.S., and Burks,C., 1985, The statistical distribution of nucleic acid similarities, Nucl.Acid Res. 13:645-656.

50. Argos,P., 1987, A sensitive tool to compare amino acid sequences, J.Mol.Biol.193:385-396.

51. Stormo,G.D. and Hartzell,G.W., 1988, Identifying protein-binding sites from unaligned DNA fragments. Proc. Nat. Acad. Sci. USA 86, 1183-1187.

52. Needleman,S.B. and Wunsch,C.D., 1970, A general method applicable to the search for similarities in amino acid sequences of two proteins, J.Mol.Biol. 48:443-453.

53. Sellers, P.H., 1979, Pattern recognition in genetic sequences, Proc. Nat. Acad. Sci. USA 76, 3041.

54. Sellers,P.H., 1984, Pattern recognition in genetic sequences by mismatch density, Bull.Math.Biol. 46:501-514.

55. Smith, T.F. and Waterman, M.S., 1981 a, Identification of common molecular subsequences, J.Mol.Biol. 147, 195 - 197.

56. Smith, T.F. and Waterman, M.S., 1981 b, Comparison of biosequences, Adv. Appl. Math. 2, 482.

57. Waterman, M.S., 1983, Sequence alignment in the neigbourhood of the optimum with general applications to dynamic programming, Proc. Nat. Acad. Sci. USA 80, 3123.

58. Waterman, M.S., 1984, General methods of sequence comparison, Bull. Math. Biol. 46, 473.

59. Chvatal,V. and Sankoff,D.,(1975), Longest common subsequences of two random sequences. J. Appl. Prob. 12,306-315.

60. Reich,J.G..,Däumler, A. and Drabsch, H., 1984. On the statistical assessment of similarities in DNA sequences. Nucl. Acids Res. 12, 5529-5543.

61. Waterman,M.S.,Gordon,L., and Arratia,R., 1987, Phase transition in sequence matches and nucleic acid structure, Proc. Nat. Acad. Sci. USA 84:1239-1243.

NEW ALGORITHMS FOR THE COMPUTATION OF EVOLUTIONARY PHYLOGENETIC TREES

Gaston H. Gonnet

Informatik E.T.H. Zürich
Switzerland

A phylogenetic tree is a representation of the evolution of species. All the phylogenetic trees are based on a model or theory of evolution. The model of evolution which we consider here is a simple model based on the similarities of peptide sequences where no parallel evolution or horizontal transfers will be considered. The vast majority of the sequence data is modern data, so we are forced to guess the structure of the phylogenetic tree. The term dendrogram is used to describe relations derived from similarities rather than from the knowledge of the evolution. In this sense our phylogenetic trees are really dendrograms.

We will assume that we do not have any additional biological information about the phylogeny; i.e. we will derive our conclusions purely from the data contained in the sequences.

The leaves of the phylogenetic tree are the present day species (or sequences from present day species). The internal nodes of the tree represent the points of divergence where two different branches of evolution arose. The root of the tree represents the nearest common ancestor of all the species considered. For example figure 1 is a phylogenetic tree where 1, 2, ..., 5 represent the present-day sequences, W represents the nearest common ancestor of 1 and 2, X is the nearest common ancestor of 4 and 5 and Z represents the closest common ancestor of all the points.

The distances of the branches between the nodes may or may not have meaning. When the distances do not have any meaning, typically all the leaves (bottom nodes) are represented at the same level. The distances of the branches between the nodes can be assigned two meanings: time or amount of evolution. Both of these measures are interesting and have useful applications. When the distances mean time, the distances between each external node and the root are identical. Properly drawn, these trees will have all the external nodes lying on a sector of a circumference. When the distances measure amount of evolution, there is no obvious constraint between them, as it is well known that similar proteins in different species may evolve at different rates.

In figure 1, if the distance means amount of evolution, we could say that 3 evolved more than 2 and 4. Since we assume that all these descended from a common ancestor, we can also say that 3 evolved more rapidly (mutated more rapidly) than 2 and 4, since in the same amount of time it evolved more.

When the prehistoric sequences (Z, Y, X and W) are not available for analysis, as it is mostly the case, we have to infer their existence and location from the present-day sequences 1, 2, ..., 5.

Computational Methods In Genome Research,
Edited by S. Suhai, Plenum Press, New York, 1994

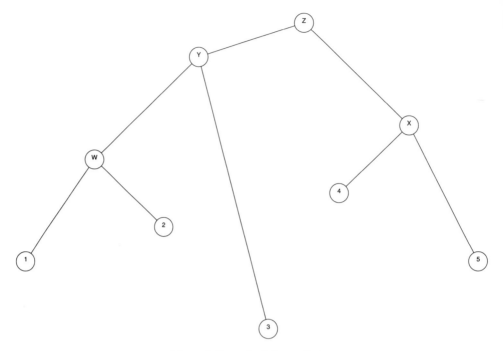

Figure 1. Example phylogenetic tree

We restrict our attention to phylogenetic trees which describe amount of evolution or evolution phylogenetic trees (ept). That is, the length of each internal branch is proportional to the amount of evolution, or more precisely proportional to the PAM distance between the nodes. The PAM distance is the most accurate measure of evolution that we can compute and ties in well with the rest of the sequence evolution modelling.

In summary, we want to reconstruct a tree, like in the above figure, for which the length of the branches are PAM distances.

The evolution between sequences can be modelled by mutation matrices. A mutation matrix, denoted by M, describes the probabilities of amino acid mutations for a given period of evolution.

Pr { amino acid $i \mapsto$ amino acid j } $= M_{ji}$

This corresponds to a model of evolution in which amino acids mutate randomly and independently from one another but according to some predefined probabilities. While simple, this is one of the best methods for modelling evolution.

A 1-PAM mutation matrix describes an amount of evolution which will change, on the average, 1 % of the amino acids. In mathematical terms this is expressed as a matrix M such that

$$\sum_{i=1}^{20} f_i \left(1 - M_{ii}\right) = 0.01$$

where f_i is the frequency of the ith amino acid. The diagonal elements of M are the probabilities that a given amino acid does not change.

If we have a probability or frequency vector p, the product Mp gives the probability vector or the expected frequency of p after a random evolution equivalent to 1-PAM unit. Or, if we start with a given amino acid (a probability vector which contains a 1 in position i and 0s in all others) M_{*i} (the ith column of M) is the corresponding probability vector after one unit of random evolution. Similarly, after k units of evolution (what is called k-PAM evolution) a frequency vector p will be changed into the frequency vector $M^k p$

Dayhoff, Schwartz and Orcutt, in their paper "A Model of Evolutionary Change in Proteins" (1978) presented a method for estimating the matrix M from the observation of 1572 accepted mutations between 34 superfamilies of closely related sequences. Their method was pioneering in the field. Nowadays we are able to estimate M by more accurate and better founded methods.

This model of evolution is symmetric, i.e. we cannot distinguish the probability of amino acid i evolving to j from the probability of j evolving to i.

The evolution from Z to Y will be modelled by a mutation matrix M^{dy}, where d_y is the PAM distance between Z and Y; the evolution from Z to 1 by $M^{dz1} = M^{dy+dw+dl}$, etc. The exponent d_{ij} denotes the distance between node i and node j. i, with a single index, denotes the distance between node i and its parent.

We reconstruct phylogenetic trees in two steps. First we approximate the topology of the tree, or the determination of the parent nodes for every internal or external node. Secondly we compute the length of the branches to fit the available data as accurately as possible. We start from the sequences and all the information we can extract from their relationships. For n sequences this information can be summarized by an $n \times n$ matrix of PAM distances, the estimate of the distance between every pair of sequences, and a similar matrix for the variances of the PAM distances.

It is important to realize that we will obtain an approximation of the original phylogeny, as all our measures are estimates. As such, several solutions, which have roughly the same probability, could be candidate answers. We are forced to decide on one candidate, the maximum likelihood candidate, the one with the highest probability of occurring. When reconstructing trees, we are potentially dealing with an exponential number of answers, so the problem of returning all answers, even with a bounded probability, is intractable.

What is new about our approach, in a quick summary, is:

- The use of PAM distances, a consistent measure of amount of evolution.
- The construction of an approximate topology and its later refinement.
- The estimation of the variance of the distances between sequences and its use in the tree construction.
- New insertion/deletion models to align sequences and estimate their PAM distance.
- New Dayhoff matrices, based on 2 orders of magnitude more data.
- Offering this service to the general public by an automatic server usable through electronic mail (to find more about this automatic server, send a message with the line `help all` to the address `cbrg@inf.ethz.ch`). This server is based on the Darwin language and system.

1. APPROXIMATE TOPOLOGY OF THE TREE

The algorithm we use to reconstruct the structure of the tree, although discovered independently, could be viewed as an extension to the algorithm by Sneath and Sokal (1973). In rough terms, this algorithm reconstructs the tree bottom up, one internal node at a time. That is, at each step one internal node, the one linking the two nodes separated by the least amount, is assigned, and from then on the two connected external nodes will be treated as a single entity. Following our example tree, lets suppose that the distances between the external nodes are approximated by the following matrix:

	1	2	3	4	5
1	—	6.4	16	17.9	19.5
2	6.4	—	15.2	17.1	18.7
3	16	15.2	—	18.3	19.9
4	17.9	17.1	18.3	—	7.2
5	19.5	18.7	19.9	7.2	—

This matrix is symmetric. The first internal node to be assigned is the one with the shortest distance, namely the one linking 1 and 2 or internal node W, since 6.4 is the minimum distance in the matrix. Now we join nodes 1 and 2 and reduce the problem, from a 5 x 5 problem into a 4 x 4. For the following steps, the external nodes 1 and 2 will behave as a single new node. Lets call this fictitious new node w. w would be like a present day sequence which averages the distance from every other node to nodes 1 and 2. So for example, $d_{14}=17.9$ and $d_{24}=17.1$ then the distance between w and 4 is:

$$d_{\omega 4} = \frac{1}{2}\left(d_{14} + d_{24}\right) = 17.5$$

and similarly for any i:

$$d_{\omega i} = \frac{1}{2}\left(d_{1i} + d_{2i}\right)$$

We have now reduced the problem by one, and obtained a new distance matrix:

	w	3	4	5
w	—	15.6	17.5	19.1
3	15.6	—	18.3	19.9
4	17.5	18.3	—	7.2
5	19.1	19.9	7.2	—

For the next step the minimum distance is between 4 and 5 with a value of 7.2. We assign a new internal node x and we perform the reduction of the matrix, averaging the distances of 4 and 5 to every other node. This results in:

	w	3	x
w	—	15.6	18.3
3	15.6	—	19.1
x	18.3	19.1	—

and now w and 3 have to be joined creating the internal node y and the matrix is reduced to a 2 x 2 matrix:

	y	x
y	—	18.7
x	18.7	—

at which point the root, Z can be assigned. Although we have constructed the tree to find its shape, we have computed an approximation of the branch lengths. At every step, we had the minimum distance between the nodes (or subtrees) to be joined, 6.4 for 1 and 2, 7.2 for 4 and 5, etc. If we assign these values divided by two to the height of the corresponding internal nodes, we obtain a first approximation of the amount of evolution in the tree. This is not insignificant, for very large samples this may be the only phylogenetic tree that we can compute. The reason is that this procedure runs in $O(n^2 \log n)$ time, whereas more refined methods run in $O(n^3)$ time. For very large families ($n > 1000$) a better approximation may be too expensive to compute.

When the variances are not equal, and/or for further steps in the computation, we have to weight the distances accordingly to compute the average distance to a group of external nodes. In general, when we have a subtree with external nodes a_1, a_2, ... and a second subtree with external nodes b_1, b_2, ... the estimate of the distance between the two subtrees based on the individual distances d_{aibj} and their variances σ^2_{aibj}

$$d_{ab} = \frac{\sum_{i,j} d_{aibj} \omega_{aibj}}{\sum_{i,j} \omega_{aibj}}$$

where we call

$$\omega_{ab} = \frac{1}{\sigma^2_{ab}}$$

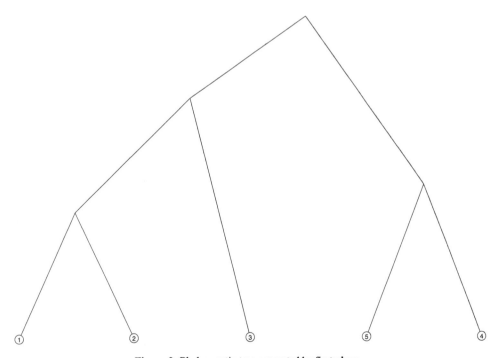

Figure 2. Phylogenetic tree generated by first phase

This formula guarantees that the contributions of each piece of information are properly weighted. All variables which we are summing for the average have their variance normalized to 1. The above formula comes from the expected value and maximum likelihood estimators for normally distributed variables. The computation can be carried by keeping the weights and the weighted sum of distances.

Computed in this way, the output will look like figure 2. It is important to notice that the topology of the tree is correct, and that the distances are somewhat accurate.

2. EVOLUTION PHYLOGENETIC TREES

Once that we have obtained the topology of the tree we can compute the length of the branches in a more accurate way. Recall that in an evolution phylogenetic tree (ept), the length of the branches are proportional to the PAM distance between the nodes. In other words, we want to build a tree such that the branch lengths conform the distance matrix as closely as possible.

Since we do not have information about the internal nodes, all our computations will be based on the external nodes (present day) information.

An ept will be one which reflects the distance constraints accurately. For example,

$$d_{y1} + d_{y3} = d_1 + d_w + d_3 \approx d_{13}$$

the distance from node 1 to node Y plus the distance from node Y to node 3 should be as close as possible to the PAM distance between the two sequences 1 and 3 In general, if A_{ij} is the closest common ancestor of the sequences i and j

$$d_{ij} - d_{iAij} - d_{iAij} \approx 0$$

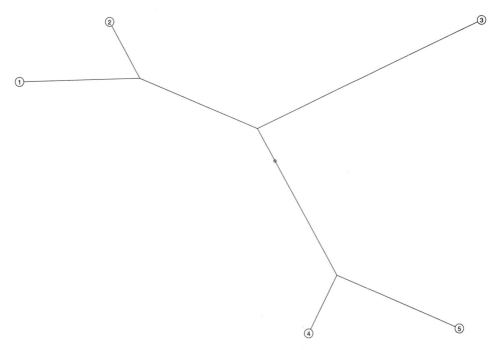

Figure 3. Phylogenetic tree generated by Darwin

The crucial observation necessary to compute the distances is that in a correct ept

$$\frac{d_{ij} - d_{iAij} - d_{jAij}}{\sigma_{ij}^2} = \varepsilon_{ij}$$

is a random variable with expected value 0 and variance 1. Optimally all the ε_{ij} would be zero, but this is not possible to guarantee since we do not have an exact measure of the distances but estimates. Then it is natural to select the internode branch lengths as the values which minimize the sum of the ε_{ij}^2. This will give us a least squares fit of the distances. This is a non-conventional least squares problem which does not present any technical difficulties except for attention to detail. The output of the above tree, as produced by our Darwin system, is shown in figure 3.

There are three sources of difficulties in the least squares computation. The first difficulty is that one pair of distances is underdetermined. The distances between the root and left and right subtrees are not completely determined. Their sum is, but the precise location of the root between these first two descendants is not. This is due to the fact that every time that one of these distances appears, it appears added with the other. This can be quickly verified in our example where all the occurrences of d_Y and d_X are of the form $d_Y + d_X$. This is a consequence of our symmetric model of evolution, and it also happens for the alignment of two individual sequences.

We resolve this difficulty with an arbitrary constraint, that the average height of all the nodes in the right and left subtree be the same. This gives a reasonable shape to the tree, but it has to be remembered that the location of the root between its two first descendants is arbitrary. To emphasize the fact that the tree is basically unrooted, we draw it as such. We call this type of representation a Larson tree. A small circle indicates where the root would be placed in the weighted centre of the tree.

The second difficulty arises when we have incomplete data and the least squares fit results underdetermined. This could happen because the all-against-all procedure may fail to align some pairs. When we do the all-against-all alignment we should only accept the matches with a similarity score above certain threshold (this threshold is usually between 80 and 100). As a result of being unable to align some sequences the tree may be disconnected. When the tree is disconnected, the connected components should be handled separately. However it may happen that the tree is connected but the number of successful alignments is not enough to resolve all internode distances. Notice that when we have n sequences, we have $n - 1$ internal nodes which give $2n - 2$ internode unknown distances. One of the two distances at the root is not determined so we have $2n - 3$ unknowns. The all-against-all matching provides at most $n(n - 1)/2$ data points. Consequently the internode distance computation

n	$2n - 3$	$n(n - 1)/2$
2	1	1
3	3	3
4	5	6

is exactly determined for $n = 2$ and $n = 3$ and overdetermined (if we have all the data points) for $n \geq 4$. On the other hand a tree could be connected with just $n - 1$ alignments which will not be enough to resolve $2n - 3$ distances.

We resolve this problem by adding a term to the least squares which minimizes the sum of the square of the lengths of the branches. This sum of squares is multiplied by a very small factor (10^{-6}), so when the problem is fully determined, this extra term will not affect the results in any significant way, but when the system of equations is underdetermined this term will force undetermined lengths to be as short as possible. This solution is entirely satisfactory in practice. It should also be noted that this sort of problem happens rarely, and only when the sequences being analyzed are marginally related.

The third difficulty is that the least squares solution may make some of the internode lengths negative. A negative length branch does not make sense in evolutionary terms. Such a situation arises from: (a) a bias in the estimation of the PAM distances, or (b) alignments which do not correspond to homology and hence the distance estimates are incorrect and (c) an incorrect topology (phase I of the construction failing). When an internode branch is assigned a negative value, this means that the least squares are not attainable at that point, but in the limit of that branch being of length zero. A branch of length 0 is not impossible, it corresponds to an ancestral sequence which split in a very short period into three descendant families (two new offsprings) instead of two (one new offspring). In other words a ternary as opposed to binary internal node. The selection of a very small value as opposed to 0 is inconsequential in the model, but allows better displaying of the tree.

Negative branch lengths do not arise frequently and only happen with data which is difficult to align correctly. It is not a serious problem in practice.

The sum of the residual squares ($\sum_{ij} \epsilon_{ij}^2$) after the fit gives a rough idea of the quality of the fit. A small number indicates a good fit, a large number indicates a poor fit. Since this value is normalized with respect to its variance, it has an expected value 1. Values lower than 1 indicate that the data was better behaved than expected. A value greater than 1 indicates that either we were unlucky or the alignments are suspicious. In any case a value greater than 1 decreases our confidence on the phylogenetic tree.

A note to the complexity of these procedures is in order. The topology construction requires a pair of matrices (double precision numbers) n x n and works in time proportional to $O(n^2 \log n)$. Thus we can use the entire memory available and expect the tree construction to complete quickly. For example with 16Mb of memory we could resolve phylogenetic trees up to 1000 sequences. The computation of ept trees has to resolve a system of linear equations of size $(2n - 2)$ x $(2n - 2)$. This requires about $12n^2$ double precision numbers in total and $O(n^3)$ time. This will restrict the size of the problems which can be solved.

3. GENERATING SEVERAL EPT TOPOLOGIES

It is always possible to improve the construction of these trees. Another improvements is to compute several likely topologies and select the one which gives the smallest total deviation. Both methods fall outside the scope of this tutorial so we will just give a brief description on how we could generate several topologies.

The step which defines the topology of the ept is the selection of the minimum distance pair in the first algorithm. As described in section 1 of this paper, this selection is done on the basis of the lowest average. When we compare two random, normally distributed, variables we have a probability greater than 1/2 of concluding the correct ordering of their expected values. But how does this probability depend on the expected values and variances? For random normal variables X_1 and X_2 distributed as N $(\bar{x}_1, \sigma_1^2)$ and N $(\bar{x}_2, \sigma_2^2)$ the probability that $\bar{x}_1$ is larger than $\bar{x}_2$ given that we sampled X_1 and X_2 is:

$$Pr\{\bar{x}_1 > \bar{x}_2\} = \frac{1 + erf\left(\frac{X_1 - X_2}{\sqrt{\sigma_1^2 + \sigma_2^2}}\right)}{2}$$

where $erf(x) = 2/\sqrt{\pi} \int_0^x e^{-t^2} dt$ is the error function commonly used in statistics. This equation shows that the probability depends on $\dfrac{X_1 - X_2}{\sqrt{\sigma_1^2 + \sigma_2^2}}$ exclusively. If we are going , allow a range of uncertainty for each comparison, this range has to be determined as :

$$\frac{\left(X_1 - X_2 \right)^2}{\sigma_1^2 + \sigma_2^2} < \delta^2$$

When X_1 and X_2 are in that range the situation is ambiguous and both orderings should be considered. When

$$X_1 \geq X_2 + \delta\sqrt{\sigma_1^2 + \sigma_2^2}$$

we assume that $\bar{x}_1$ is definitively larger than $\bar{x}_2$ and similarly for the symmetric case. is δ our tolerance parameter. The larger δ is, the more confidence we will have in not making mistakes on the tree topology. Also, as δ increases we may have to analyze more and more topologies.

At each node-joining step in the topological tree construction we may have to consider several candidates to join. When the shortest distance is given by the pair i, j, then all the other pairs k, l satisfying

$$\frac{\left(d_{ij} - d_{kl} \right)^2}{\sigma_{ij}^2 + \sigma_{kl}^2} < \delta^2$$

and either

$i = k$ or $j=k$ or $i=l$ or $j=l$

should be considered. Each one of the pairs will generate a different topology. The second condition is necessary because the merging order of two pairs with no overlap will not alter the resulting tree. When we are left with $m > 1$ pairs of distances we have to create one tree for each of the pairs. The procedure continues for each of the pairs independently. This will multiply the number of trees by m. It is easily seen that this procedure will create an exponential number of trees, and hence δ should be set judiciously low. The lower that δ is, the fewer pairs will have to be considered as overlapping and the lower m will be.

Once we have found a reasonable number of different topologies, say 5 to 20, we can compute the least squares approximation for each tree and select the best one. Alternatively we could inspect these trees and try to conclude information from the different shapes which arose.

MODELLING PROTEIN STRUCTURE FROM REMOTE SEQUENCE SIMILARITY: AN APPROACH TO TERTIARY STRUCTURE PREDICTION

William R. Taylor

Laboratory of Mathematical Biology
National Institute for Medical Research
The Ridgeway, Mill Hill
London NW7 1AA
UK

1 INTRODUCTION

Given some similarity in sequence between two proteins, a molecular model can be constructed for one sequence provided that the tertiary structure of the other is known. This approach, which is often called "modelling by homology", can yield a reasonably accurate model, providing that the available sequence similarity is clear. With less similarity, useful models can still be constructed and, although these may not be accurate in detail, it is probable that the overall fold will be correct. The robust nature of the approach derives from the stability of the fold of the protein under mutational pressures. This is so conservative that, for two proteins, sequence similarity can be almost undetectable yet the overall fold can remain the same. By implication, if any reliable sequence similarity can be identified at all, then it should be possible to build, at least a rough, model with the correct fold which may well be reasonably accurate in detail in the core.

For these rough models, the homologous modelling approach is limited only by the extent to which two sequences can be reliably aligned. With only two sequences, however, there is a limit to the degree of similarity between them that constitutes a meaningful alignment and additional sources of information are needed to go beyond this. These can come both from consideration of the tertiary structure and from the multiple alignment of additional related sequences. This latter aspect is most useful when a subfamily can be gathered around both the sequence of known structure and the sequence to be modelled. In this situation the patterns of conservation in each subfamily can be equated allowing a more accurate alignment to be obtained.

This improved situation for modelling a new sequence also has its limit which is now determined not by a simple statistical measure but depends on the degree of similarity of the sequences that constitute each subfamily. If the sequences within both families are too closely related then the situation is equivalent to having only a pair of sequences but if they are too distantly related then the alignment within each subfamily may be incorrect. There is therefore an optimal situation with several remotely related the sequences in each subfamily and, in general, the more distant are the relationships in each subfamily, so the more remote will be the similarities that can be established between the two families.

To obtain an accurate alignment of remotely related sequences is often difficult but if both have a known structure, then the sequences can virtually be ignored and the structures themselves compared directly. However, even with the explicit information available in the structures, this is still not a simple task when the protein structures are distantly related. Commonly, equivalent loop regions will have different lengths and substructures may have shifted relative to each other.

In the most favourable modelling situation there might be six sequences in each subfamily, with, at least, two of known structure in one of the families. Two basic computational tools are then required to establish an alignment in each of the subfamilies. An accurate (and preferably fast) multiple sequence alignment and a flexible structure comparison program. An optimal strategy would then be to determine an alignment based on the known structures and against this reference, align any further sequences in that subfamily. The second subfamily, which has no protein of known structure must be aligned on a sequence basis alone. The final step of aligning the two alignments will depend on the degree of residual similarity between the subfamilies and may be accomplished by reference to common patterns of conservation, interpreted in the light of the known structures.

An example of model building from very remote sequence similarity, using all the techniques discussed above, can be found in the construction of a model for the retroviral (HIV) proteases [1]. This model, which was based on the structures of the aspartyl proteases, was substantially verified by later X-ray studies on both a related virus and HIV itself. However, because the protein functions as a dimer, the structure of the β-strands at the dimer interface was not correct. In the X-ray structure, these form an interlocked structure while the dimeric model was generated by a simple juxtaposition of the intact monomer across the two-

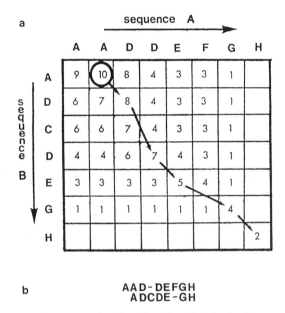

Figure 1. The basic Dynamic Programming algorithm: Steps in this algorithm are illustrated using the alignment of two short sequences: A and B. Matching positions score 2 and the insertion of a gap costs 1. a — Beginning at the ends of the sequences (bottom right of the matrix) the score of 2 is entered in the cell corresponding to the matching Hs. Sums of the scores are accumulated working back towards the start of the sequences. Thus, the cell corresponding to the aligned Gs scores 2 and inherits 2 from the matched Hs, giving 4. None of the other positions on the same row or column have any matches but they still inherit 2 from the matched Hs less 1 for the insertion of a gap (giving rise to lines of 1s). The next diagonal position (EF match) has no score but inherits the preceding score of 4. However an insert of a gap allows Es to match gaining 1 (2 less the gap penalty). This is continued until the matrix is complete. The highest score is then found (circled 10) and its inheritance path is traced back to the opposite corner of the matrix (heavy arrows). b — The resulting alignment. This simple process has the remarkable property that it guarantees to find the best alignment for the given constraints.

fold axis. [2] gives a summary of these changes while [3] provides a more detailed analysis of the predicted model and one built from a closely homologous X-ray structure.

The following discussion will review the techniques which form the basis of the structure and sequence alignment tools outlined above and then progress to some of the more novel approaches that can be applied to very remotely related sequences which have only partial similarity. Finally, the ultimate extension of these techniques to problems where no specific similarity is found will be discussed.

2 ALIGNMENT TOOLS

Proteins are linear polymers and this fundamental property provides a powerful constraint which can be utilised in the algorithms for their comparison. In other words: the algorithms need consider only relative insertions and deletions of sequence while transpositions and inversions can be ignored. This constraint is used by the Dynamic Programming algorithm to find an optimal alignment between two sequences, given a numeric measure relating their components (bases or residues). The same algorithm can be used in the comparison of a pair of sequences and a pair of structures and even for the comparison of a sequence to a structure. Because of this great generality, the basic algorithm is outlined in Figure 1.

2.1 Multiple sequence comparison

The majority of multiple alignment methods currently in use, employ a simple heuristic that it is best to align the more related pairs of sequences first, and then further combine these sub-alignments. When this hierarchic condensation is combined with averaging over the aligned sequences, then the method becomes more robust since the conserved regions remain coherent and eventually dominate the alignment. When used in combination with a pre-filter based on peptide composition the latter approach can be used with thousands of sequences.

2.2 Structure Comparison

The problem of comparing three-dimensional protein structures has generally been based on the superposition of the structures as rigid bodies. While this is sufficient for closely related structures, more remotely related structures often need to be broken into fragments. This solution has the problem that the relationship between the fragments is lost in the calculation which makes the approach inaccurate for remotely related structures. A different approach, pioneered independently by [4] and [5] considers instead the similarity in the structural environment of residues in each protein. This measure has the great advantage that the Dynamic Programming algorithm can be used, allowing insertions and deletion of sequence to be easily handled. The computational problem in 3D is much more difficult than the equivalent (1D) sequence alignment problem but can be solved with the help of stochastic minimisation methods or by the double application of the Dynamic Programming algorithm. The latter solution is rather more rigorous and utilises the additional constraint of the linear nature of the protein sequence throughout. (Figure 2).

The solution of the problem for pairwise data allows any property that can be expressed in this way to be used to constrain the alignment and can include hydrogen-bonds or disulphide bonds. The use of the Dynamic Programming algorithm also allows the trivial incorporation of any properties that are intrinsic to a single amino acid, such as solvent accessibility, torsion angles and, of course, the chemical nature or identity of the amino acid itself. Thus sequence and structural alignment can be performed simultaneously. Further developments of the method of [5] include a visualisation method, a fast version of the algorithm, including a version that compares secondary structures, so providing a rapid pre-filter to the more detailed (residue level) calculation. A local alignment version based on the [6] algorithm has been developed, along with a multiple version based on the algorithm of [7].

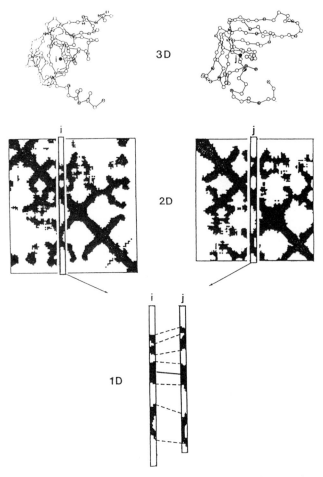

Figure 2. Outline of the protein structure comparison method of Taylor and Orengo (see text for details). The method employs a core algorithm normally used in sequence alignment. To use this algorithm on three dimensional (3D) protein structures *(a)*, these are first reduced to a two dimensional representation in the form of a distance plot *(b)*. From the two plots, pairs of columns (1D) are compared *(c)* as if they were protein sequences using the Dynamic Programming algorithm (Figure 1). Since each column only contains a fraction of the data, all pairs of columns *(i,j)* are compared and a consensus alignment derived.

3 SEQUENCE TO STRUCTURE COMPARISON

Using the same algorithm for both sequence and structure comparison allows both aspects to be combined producing a hybrid method for the comparison of a sequence to a structure. Given a protein structure (ignoring its own sequence), each residue position can be considered as a blank position on which a residue of the new sequence can be located. For each alignment or threading of the sequence onto the structure it is possible to assess how well the resulting model has packed the hydrophobic amino acids and exposed the polar amino acids to solvent. Because there are very mans that a simplified model must be used. We therefore adopted empirical pseudo-energy potentials derived from an analysis of the protein structure data bank. These provide a measure of how favourable it is to find any pair of amino acids at a given distance. For each threading, the sum of these 'energies' over all pairs of residues gives a defined measure for which a minimum can be found. The threading with the minimum energy is then the predicted alignment of the sequence onto the structure (Figure 3).

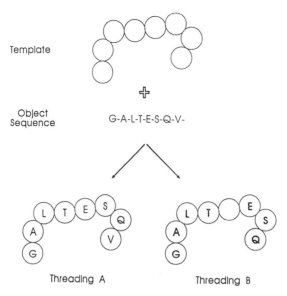

Figure 3. Conceptual view of sequence threading. The *TEMPLATE* is a protein structure considered only as a blank series of spatial positions (ignoring its amino acid sequence). Onto this a new sequence is *threaded* by trying different allocations of amino acids to locations. Each possibility (eg: A and B) is evaluated by rough empirical pairwise energy potentials. An exhaustive cominatoric search is avoided by using the *Double Dynamic Programming* algorithm (see text for explanation) allowing optimal solutions to be found for large proteins in a reasonable time.

The calculation of all possible threadings appears initially to be an impossibly large (combinatoric) search problem. However the estimated distances are pairwise data and can therefore be used in the structure comparison (or Double Dynamic Programming) algorithm of [5]. This can be viewed intuitively as replacing one of the distance plots (Figure 2) with predicted rather than observed distances. The algorithm will then proceed as before, irrespective of the source of the data found in the matrices.

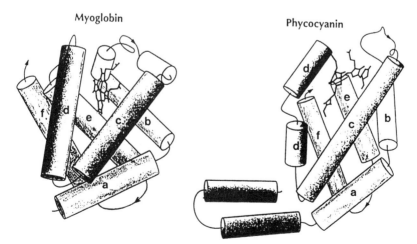

Figure 4. The sequence of phycocyanin (right) was recognised as being similar to that of a globin (left) when matched against the protein structure databank using the *threading* method. There is no significant sequence similarity between the two proteins yet they have the same fold (α-helices are represented by cylinders and the chain direction marked by arrow heads). Phycocyanin has two additional helices at the N-terminus which have no correspondence in the globin fold and these were correctly ignored by the program giving the correct equivalence of secondary structures (identified by corresponding letters on helices).

Similar methods have been recently developed by others, but these are more closely related to simple sequence alignment whereas our approach compares a sequence directly to a structure. Although the method is still under development and assessment, it appears to be capable of recognising relationships between proteins that could not be detected by any method using sequence information alone. A particular example is the ability of the method to find the globin fold in the phycocyanin sequence: even with multiply aligned sequences, this relationship is not apparent. This then opens the opportunity to pursue the approach of modelling by homology to the most remote limits. (Figure 4). As with the structure comparison problem, the technique is being extended to allow local comparisons and multiple sequence (and structure) data.

4 TOWARDS TERTIARY STRUCTURE PREDICTION

A limitation of the threading and modelling methods outlined above, is that they can only succeed if the sequence of unknown structure is related to a protein of known structure in the structure data bank. Occasional attempts are made to predict a structure directly from sequence

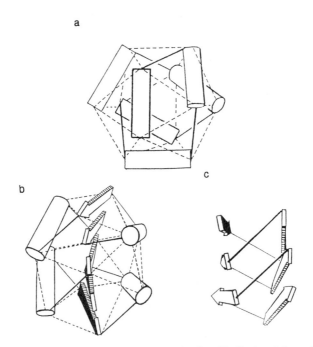

Figure 5. *a* — By assuming α-helices have equidistant end-points, idealised models can be contracted for small globular proteins. Six helices have 12 ends which define an icosahedron (less helices define less regular polyhedra). All windings over this surface define a protein fold which can then be investigated. *b* — β-strands can be incorporated into this simple framework by allowing them to lie at half the helical spacing. End-points were defined for a small β/α protein by connecting two hexagons that have a relative rotation of 30°. This imparts a suitable twist to the sheet (arrows) and a corresponding twist between both strands and helices. *b* — The model can also be used for stacked β-sheet proteins by neglecting the β-strands in *b*, reducing the scale by half and taking the α-helices as β-strands. With a slight shift, both *beta/beta* and *beta/alpha* models can be extended into a super-helical structure, so allowing any number of β-strands to be incorporated. A simple winding is shown which is not typical for this class of protein (they almost exclusively have antiparallel strands in the sheet).

by packing secondary structures which have usually been identified from multiple sequence alignments. However, despite, often reasonable secondary structure predictions, no novel fold has ever been correctly predicted for anything other than small proteins where the packing possibilities are very limited.

An alternative strategy is outlined below which attempts to circumvent the problem of predicting large structures. This approach aims to generate a large number of plausible folds based on simplified models (or architectures) against which the new sequence can be compared using the threading method (described above). Although the idealised structures may not match the unknown structure exactly, it is hoped that the flexibility inherent in the threading method will allow a match to be found.

4.1 Generating plausible folds

For a given class of globular protein, a large number of folds can be generated by a combinatorial procedure. This begins by selecting a suitable framework composed of line segments (sticks) representing secondary structure locations. The model for α-helical proteins is that of [8] which has been adapted for both the β-α class and β sandwich type of protein (Figure 5). In these models the edges (sticks) are simply place-holders for the secondary structures and are of equal length, undirected and unconnected. Following a combinatorial tree search, every winding of the sequence over the sticks is generated, subject to any local rules which restrict handiness of connection or crossing of loops.

Each fold can then be ranked on the basis of general principles of protein structure, and if some local substructure (motif) has been identified in the sequence, then only those folds that accommodate it need be selected. For well defined motifs, such as the calcium binding EF-hand structure, the selective power is considerable. For example; for the six α-helices in parvalbumin there are 1264 possible folds, but only three can accommodate the double EF-hand motifs (Figure 6) [9].

a

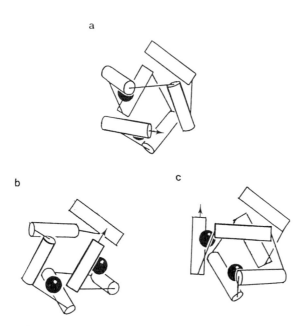

b   c

Figure 6. The all-α model (Figure 4a) accommodates 1264 different folds for six helices. Those that have the final two pairs of helices in the conformation of an EF-hand (calcium binding motif) are potential models for parvalbumin. Only three such folds exist which includes a close approximaiton of the native parvalbumin fold (*a*). The two alternatives differ only in the orientation of the first two helices. Bound calcium ions are represented as black spheres, α-helices as cylinders, and the chain direction by an arrow-head on the carboxy-terminus.

4.2 Stick model to residue model

Each secondary structure based fold which bears some resemblance to a globular protein must be rendered in sufficient detail to allow the threading algorithm to be applied. This requires that each secondary structure axis is elaborated into an α-carbon position for each residue. Where the secondary structure has an appreciable hydrophobic moment, then this can simply be oriented towards the centre of gravity to provide a good starting model for further refinement.

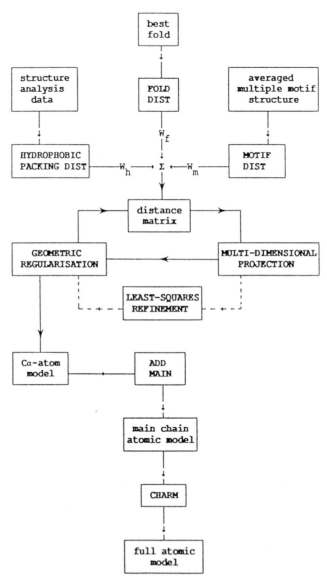

Figure 7. Outline of methods to convert a secondary structure "stick" model to a realistic protein structure. The initial model is specified by a linear (weighted) average of pairwise distances derived from the stick model (FOLD DIST.), specific motif distances (MOTIF DIST.) and a generic propensity for conserved hydrophobic positions to be close (HYDROPHOBIC PACKING DIST.). The resulting matrix is projected into real-(3D)-space and regularised to restore bond lenghts and steric repulsion. The option for fuller least -squares refinement is computationally slow and seldom used. The final α-carbon model can then be elaborated into a full atomic model using conventional methods.

Typically, of the order of 100 folds might be selected as potential candidates for further examination at the residue level. This requires that the refinement method must be sufficiently fast to construct this number of models in a reasonable time. To achieve this, a hybrid form of Distance Geometry and fast real-space geometrical regularisation has been developed which can construct a protein model of 100 residues in 60 seconds on a typical workstation (Figure 7). The resulting models have well packed and buried hydrophobic residues and realistic loop conformations and by 'eye' are not apparently different from their less fictitious relatives in the structure data bank [10].

4.3 Threading predicted folds

Since the threading algorithm does not use side-chain locations in its empirical energy function, each fold generated by the preceding methods can simply be treated as if it were a 'real' data bank entry. However, given that systematic deviations will appear in the predicted folds, it is unlikely that the results of this application will be as reliable as finding a match in the structure data bank. Preliminary results indicate that the approach may isolate a limited number of folds that can then be assessed in greater detail or be discriminated on the basis of biochemical or NMR data.

REFERENCES

1. Taylor, W. R., Orengo, C. A., and Pearl, L. H. (1990). Comparison of predicted and X -ray crystal structures of retroviral proteases. In Ikehara, M., editor, Protein engineering: protein design in basic research, medicine, and industry, pages 21-27. Japan Scientific Press, Tokyo; Springer-Verlag, Berlin. Proceedings of the second international conference on protein engineering.
2. Weber, I. T. (1990). Evaluation of homology modeling of HIV preotease. Prot. Struct. Funct. Genet. 7:172--184.
3. Pearl, L. H. and Taylor, W. R. (1987). A structural model for the retroviral proteases. Nature, 329:351-354.
4. Sali, A. and Blundell, T. L. (1990). Definition of general topological equivalence in protein structures: a procedure involving comparison of properties and relationship through simulated annealing and dynamic programming. J. Mol. Biol. , 212:403-428.
5. Taylor, W. R. and Orengo, C. A. (1989b). Protein structure alignment. J. Molec. Biol. , 208:1-22.
6. Smith, T. F. and Waterman, M. S. (1981). Comparison of bio-sequences. Adv. Appl. Math. , 2:482-489.
7. Taylor, W. R. (1988). A flexible method to align large numbers of biological sequences. J. Molec. Evol., 28:161-169.
8. Murzin, A. G. and Finkelstein, A. V. (1988). General architecture of the α-helical globule. J. Mol. Biol. 204:749-769.
9. Taylor, W.R. (1991) Towards protein tertiary fold prediction using distance and motif constraints. Prot. Engng. 4: 853-870.
10. Taylor, W.R. (1993) Protein fold refinement: building models from idealised folds using motif constraints and multiple sequence data. Prot. Engng. 6: 593-604.

GENETIC ALGORITHMS IN PROTEIN STRUCTURE PREDICTION

Frank Herrmann and Sándor Suhai

Department of Molecular Biophysics
German Cancer Research Center
Im Neuenheimer Feld 280
D-69120 Heidelberg, Germany

ABSTRACT

Genetic Algorithms are optimization techniques. In protein structure prediction, tentative structures are varied and evaluated according to a certain rate of assessment. The rates are optimized in order to achieve a prediction of the real structure. A typical approach is to minimize the conformational energy using an empirical force field. The optimization method is faced with two difficulties. First, the space of possible conformations is very large even for small proteins. Second, the objective function usually is highdimensional and multimodal: this is known as the multiple minima problem. Genetic Algorithms are promising to be less limited by these problems than common optimization methods. They try to exploit the mechanisms by which natural evolution performs its optimization task - to create life that is optimally adapted to its environment.

1 INTRODUCTION

First to avoid a common misunderstanding: Genetic Algorithms are not evolving programs. All Genetic Algorithms are variations of one prototype algorithm. The actual algorithm chosen for a particular optimizaton task is fixed during an optimization run. The name Genetic Algorithm stems from the method of operation of the prototype algorithm while it simulates a simplified model of natural genetics. The evolving parts of the Genetic Algorithm are the solutions to the optimization task that are manipulated during the optimization run.

The parts of the prototype algorithm and the performed actions during an optimization run are explained in the first section. Then the interface to the optimization task and how to link the user supplied problem to specific parts of the prototype algorithm are described. Finally, the mechanisms by which the optimization makes progress are accounted for. The second section gives a brief history of the origins of Genetic Algorithms. The third section supplies an overview of Genetic Algorithm applications in different fields. The fourth section focuses on protein structure prediction. It discusses existing applications and compares them to traditional methods of protein structure prediction. The fifth section proposes an extended approach to protein structure prediction using Genetic Algorithms. The last section gives references for those who want to make own experiments.

Computational Methods In Genome Research,
Edited by S. Suhai, Plenum Press, New York, 1994

2 METHODOLOGY

Genetic Algorithms contain a population of candidate solutions that are encoded as binary strings. A sequence of generations is generated by building new populations from the previous ones. The members of the new population are selected from the old one according to their objective values. The better the value, the greater the fitness of the member and the probability that it occurs in the next population. After this stage of selection, the genetic operators are applied. The genetic operators introduce the mechanisms of inheritance. The most important operators are crossover and mutation. The crossover combines parts of two parent solutions to a new solution. The mutation randomly changes the binary representation of a solution. Figure 1 gives a schematic presentation of this procedure. The procedure is iterated for each generation until a termination criterion is fulfilled. Figure 2 presents the flow-chart for the main loop of a Genetic Algorithm. The selection, combined with the genetic operators, generates populations with increasing best objective values.

2.1 Selection

The selection chooses members of the population with a certain selection probability. This probability depends on the fitness of the member relative to the fitness values of the other members. A member whose fitness exceeds the average fitness of the population has a selection probability that makes it likely that it will have more than one ancestors in the next generation. The fitness value is computed from the objective value. According to the goal of

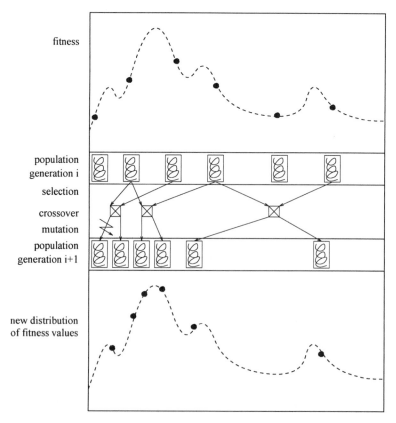

Figure 1. The steps performed during one generation

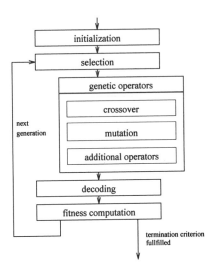

Figure 2. The iterated main loop generates a sequence of populations

the optimization, minimization or maximization , the objective values are mapped to fitness values, so that better objective values have greater fitness values. The selection probability p_i is computed from the n fitness values f_i by means of the formula:

$$p_i = \frac{f_i}{\sum_{j=1}^{n} f_i}$$

2.2 Genetic Operators

The crossover operator generates two new strings from two old strings. It swaps a part of the bit string that is determined by a crossover point. This crossover point is chosen randomly and the part of the string at the right side of it is exchanged between the two strings.

Example:
$$Crossover\{101011, 111001\}:$$
$$101011 \longrightarrow 10|1011 \longrightarrow 101001$$
$$111001 \longrightarrow 11|1001 \longrightarrow 111011$$

The mutation operator randomly chooses a bit position in the string. Then it flips this bit. The mutation operator is able to generate strings that cannot be generated by the crossover, if a bit has the same value in the entire population.

2.3 The Interface to Applications

Genetic Algorithms can be applied to many problems. There is an interface between the problem independent core of the Genetic Algorithm and the user supplied problem specific part that varies at different applications.

The Genetic Algorithm manipulates binary strings. The coding determines which are the corresponding solutions to the optimization task. After each application of the genetic operators, the largely new binary strings are decoded (See Figure 2). The resulting candidate solutions are evaluated by a user supplied function that assigns objective values measuring the quality of the solutions. The objective values are then mapped to fitness values that in turn

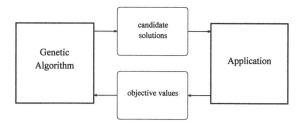

Figure 3: Interface between GA and Application

are used by the Genetic Algorithm to generate the next generation (See Figure 3). The user only has to supply the coding and the objective function that judges proposed solutions but he does not have to specify a process to generate tentative solutions. By using binary strings, the Genetic Algorithm is able to perform the same procedure to generate new solutions regardless of the entities to be optimized.

2.4 Why Genetic Algorithms work

On the first glance, it seems as if Genetic Algorithms perform a random search. They don't because the search space isn't sampled with uniform probability. The schema theory provides evidence that the search is directed towards better solutions. The Genetic Algorithm implicitly detects correlations between high fitness values and patterns in the binary strings. Good solutions exhibit similarities in their binary strings that correspond to building blocks of better solutions. They define promising subspaces of the search space that will be sampled more frequently.

2.5 Advantages of Genetic Algorithms

Genetic Algorithms have several advantages in comparison with traditional optimization techniques. First, one does not have to supply a procedure how to interpret previously calculated search points to generate new search points with probably better objective values. One only has to supply the objective function and the rest will be done by the Genetic Algorithm.

Second, the objective functions are allowed to have multiple local extrema, to be highdimensional and discontinuous. Traditional optimization techniques often require properties that most of real world problems lack. With gradient methods, for instance, the search of a multiple extrema function terminates in the local extrema next to the starting point. In contrast hereof, Genetic Algorithms are global search methods.

2.6 Genetic Algorithms as Simulated Evolution

Genetic Algorithms simulate a simplified model of natural evolution. The population of candidate solutions corresponds to a population of creatures. The binary strings are the chromosomes and when two creatures are crossed, they create their offspring that inherits parts of their genes. Mutation simulates errors during the reproduction or other random changes of the chromosome. According to the survival of the fittest, the solutions with the best fitness values reproduce most and are more likely to produce offspring.

3 HISTORY

The first simulations of evolution were motivated by biology. Formalized models of natural evolution were employed to reproduce observed processes in nature. Population genetics uses mathematical models to describe natural evolution [1].

Later on, simulations with a different goal appeared. Inspired by the success of natural evolution in generating adaptations to complex environments, Genetic Algorithms were used to solve complex optimization tasks by means of artificial evolution. The founder of this approach is Holland [2]. Goldbergs book [3] marks the beginning of a renewed interest, also indicated by international conferences and the news group of Genetic Algorithms.

The first applications of artificial evolution based on a model of evolution that stresses the importance of mutation as the key to an ongoing adaptation. Evolution was seen as a trial and error process and only the large time spans of natural evolution were thought to achieve the wide variety of complex organisms.

These models had considerable success as optimization procedures for engineering purposes [4]. Successful applications comprise the design of streamlined bodies. A single candidate solution was mutated at one location at a time and each deterioration was decarded. But for more sophisticated problems this proved to be a too large simplification of natural evolution. It turned out that it's crucial to use a whole population of candidate solutions and that the main driving force of a rapid adaptation is the recombination carried out by the crossover operator.

4 APPLICATIONS IN MATHEMATICS

4.1 Function Optimization

A straightforward application of Genetic Algorithms is function optimization. The strings are vectors of the function arguments and the resulting function value is the objective value. The arguments are binary coded, in normal or gray code, for instance.

De Jong [5] used a test environment of five functions with different properties. He investigated the performance of Genetic Algorithms on these functions and the effect of diverse settings of the Genetic Algorithm parameters. These are the population size, the crossover, and mutation probabilities. Figure 4 shows typical curves of the best value and the population average of the objective value in a minimization task. De Jongs simple multidimensional quadratic function f1 was used with 20 dimensions and a population size of 20.

$$f_1(x_i) = \sum_1^{20} x_i^2, \qquad -5.12 \le x_i \le 5.12$$

Ackley [6] compared 7 optimization algorithms in an empirical study. Two of them are variants of a Genetic Algorithm using uniform and ordered crossover, respectively. The other algorithms comprise variants of hillclimbing, simulated annealing and a combination of stochastic hillclimbing and a Genetic Algorithm. Tests are performed using a test suite of 6 functions with a bit vector as argument. The following enumeration quotes the different properties of the functions given in braces: one maximum (OneMax), two maxima (TwoMax), a global and a local maximum with a large basin of attraction (Trap), a shape like a porcupine (Porcupine), several plateaus (Plateaus) and a mixture of this all (Mix), respectively. The tests are performed with bit vector lengths of 8,12,16 and 20 except for the mixed function, where they are 15,20,25 and 30. Each combination of algorithm with function with bit vector length is tested 50 times.

The properties of the functions refer to the notion of distance of the argument from the optimum. That means, the properties describe how the function values change, if the arguments escape from the optimum. The optimum is always the bit vector with all bits set to 1. The distance in the multidimensional binary space is calculated by counting the bit matches between the actual argument and the optimum. The function value is calculated from the match count and the bit vector length.

f1(x)

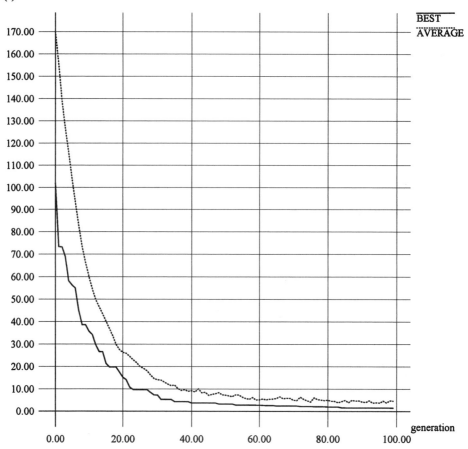

Figure 4. The best and the population average objective value is shown for each generation. Here the objective value is minimized.

The functions are chosen to represent stereotypical properties of search spaces. The properties depend on the defined distance measure. If one interprets the employed coding as model for multidimensional parameter optimization then one bit corresponds to the parameter of one dimension and the distance from the optimum is the sum of the distances in each dimension. The single bit can only distinguish between no distance and maximal distance in the corresponding dimension. Furthermore the effects inside the coding of one dimension are neglected. If interpreted as model for a onedimensional parameter optimization then each bit uniformly contributes to the distance whereas in common binary codings the bits have different valences.

The performance of the algorithms is measured by the number of function evaluations executed before reaching the optimum. The values of the 50 runs of each combination are averaged to compare the results. A maximum of 1 million evaluations is used to skip the remaining tests of a combination since it is not guaranteed that the optimum is always reached. The algorithms are compared separately for each function by ranking them according to their performance. The following enumeration quotes the ranks of the two versions of Genetic Algorithms for the function given in braces. The numbers denote the rank out of 7 ranks with rank 1 the best one: 4,5(OneMax), 4,5(TwoMax), 6,7(Trap), 2,3(Porcupine), 2,4(Plateaus), 2,5(Mix). The function with the local maximum with the large basin of attraction exceeded the maximum of evaluations for the 16 and 20 bit vector lengths. This is approximately the size of the search space. An optimization performed with the GENESIS system likewise using a population size of 50, a bit vector length of 20 and 50 tests resulted in a minimum of 203, a maximum of 408 and an average of just 314 evaluations. Therefore, Genetic Algorithm's are not generally trapped by large local optima. The Genetic Algorithm employed in the study shows several differences to a standard one, it uses a threshold for selecting the next population rather than proportional selection, for instance. It may be due to these differences that the GENESIS system performs better on the function Trap.

4.2 Traveling Salesman

The Traveling Salesman Problem is to find the shortest tour that visits all given cities and returns to the start city. The computational effort grows exponentially with the number of cities, so an enumerative search isn't affordable for more than a few cities.

This application shows two issues that have to be considered carefully by users of Genetic Algorithms. First, the crossover must generate only legal ancestors for all possible parents and crossover points. Second, the coding should support the formation of partial solutions. That means that similarities in the strings correspond to similarities in the entities to be optimized and, therefore, correlate with good fitness.

One alternative representation of a tour uses the sequence of the numbered cities. Though a part of a sequence represents a part of the tour but after crossover tours may have duplicates of some cities while others will be missing.

Another representation uses indices into a list of cities still to be visited. This way equal numbers in a tour denote different cities and cities cannot occur twice or be omitted. But parts of a sequence mostly represent completely different partial tours depending on the prefix of the sequence. So the search cannot exploit correlations between parts of the strings and good fitness and, therefore, it performs somewhat of a random search.

To circumvent these problems, one can use knowledge-augmented operators. Problem specific knowledge is incorporated into the operators. For the Traveling Salesman Problem a crossover operator is used that generates legal tours from two parent tours [7]. This application obtained good results with a computational effort comparable to simulated annealing.

5 APPLICATIONS IN STRUCTURE PREDICTION

There are only a few recent publications of molecular structure prediction using Genetic Algorithms. This indicates that the field is still at its rise. In this section an overview of existing applications is provided. The next section suggests an extended approach to protein structure prediction using Genetic Algorithms and mentions results already obtained with this approach.

5.1 Generating Structures from *NMR* Data

Blommers et al [8] used a Genetic Algorithm to obtain structures of a dinucleotide that best satisfy two sets of *NMR* constraints. The first set is used to define the search space and the second set to rate the samples. The generated structures are further refined by energy minimization using a steepest descent and subsequently a conjugate gradient method. The *NMR* constraints comprise distance restraints and bounded torsion angle regions. The torsion angle regions define the space of possible structures. Hence, the population was initialized by randomly selecting dihedral angles from the torsion angle regions. The Genetic Algorithm generates conformations that were tested against the distance restraints and were assigned a corresponding fitness value. A specific selection scheme called sharing was used to allow for remaining genetic diversity. In this way different conformations converge side by side and multiple starting points for energy minimization were obtained. This is meaningful since it could be inferred from the nmr constraints that the molecule is not conformationally pure and that there are more than one conformations in solution. The inconsistent nmr data that indicate two dominant conformations are split into sets that agree with one of these, respectively. The Genetic Algorithm calculations are performed using a population size of 100. After 25 generations, conformations occurred that violate none of the distance constraints. After refinement with molecular mechanics, one of the two generated conformations is identified as the major conformation by its lower energy. The conformations were superimposed to distinguish the flexible and the rigid parts of the conformation. A part of the conformation were compared to crystallographic data from a strongly related molecule. The torsion angles were seen to be in good agreement. The usefulness of Genetic Algorithms in structure determination is judged in comparison to a common procedure that generates samples by distance geometry refined by molecular mechanics. The success depends on a proper sampling of the conformational space. The Genetic Algorithm exhibited an explorative sampling during the first generations and an exhaustive sampling around the local minima at the end.

5.2 Optimizing Protein Side Chain Conformations

Tuffery et al [9] have chosen the backbone of the protein fixed and optimized the corresponding side chain conformations. The results are checked against data from crystal structures from which the backbone is taken. So called rotamers are used instead of all possible dihedral angles. They are combinations of side chain dihedrals that frequently occur in known protein structures. The *Protein Data Bank* is used to assemble a set of structures with resolution better than 2 Å and with numbers of amino acids ranging from 36 to 437. A dynamic cluster analysis was performed and resulted in a set of 109 side chain conformations. The application of rotamers significantly reduces the search space. The objective value is the conformational energy calculated by the FLEX force field. The underlying theory is that folding protein molecules perform an energy minimization and, therefore, reach the minimum in their native conformation.

The used modification of a Genetic Algorithm called Selection Mutation Focusing Algorithm (SMF) heavily differs from a Genetic Algorithm as defined by Goldberg [3]. The selection, the reproduction and the crossover operator were implemented in a different manner. Instead of using proportional selection, the new population is generated by applying an energy threshold and subsequently filling up to the original population size. The new members are generated according to a *probability law of reproduction* , that means that they are randomly generated by choosing the rotamers according to probabilities obtained from a convergence analysis of the selected members. It was observed for this kind of Genetic Algorithm, that "in general the crossover operator has a minor influence on the search", whereas in real Genetic Algorithms it is the main driving force of search.

Optimizing the side chains has the advantage that the side chains mainly interact with their closest neighbours. So distant side chains can form local partial solutions that don't depend on each other. Moreover, the heavy interaction of sidechain and backbone conformation does not complicate the optimization. Besides the Genetic Algorithm, a heuristic method were also used. The heuristic Sparse Matrix Driven Algorithm (SMD) makes use of the weak interactions of side chains wide apart. The problem is split into subproblems governing strong side chain-side chain interactions, for which a full combinatorial search is performed. The SMD appeared to be preferable with respect to computational effort when a single solution is required. The Genetic Algorithm leads to a set of quasi-optimal conformations which can serve as starting points for molecular dynamics.

5.3 Modifying Wild Type Sequences

Dandekar and Argos [10] applied Genetic Algorithms to three different tasks, two protein engineering and one protein folding application. First, the space of primary structures is searched for potential zinc finger sequences. Second, a wild type λ repressor sequence is altered corresponding to user specified requirements. Third, an ab initio fold of a protein is modelled. The first application uses a population of 100 strings coding the 30 residues of the potential zinc finger motif with 180 bits. The optimizations are repeated up to 24 times each with a different randomly initialized population. The best solutions of these so called epochs were used to initialize the start population of a final optimization. The fitness function measures if certain properties of known zinc finger sequences are present. It sums weighted measures of amino acid composition and matches of amino acids at certain positions with allowed amino acid types from a zinc finger consensus sequence. The Genetic Algorithm generated zinc finger like sequences within 700 generations.

The second application produces hypothetically more stable derivatives of the λ-repressor sequence. A Genetic Algorithm is used to establish user specified properties while conserving properties of the wild type. Residues are exchanged with those that showed a greater preference for the structural type. The fitness function includes terms for solvent accessibility, the volume of core sidechains, amino acid preferences for turns and helices and the occurrence of user preferred amino acids. The user can change the weights associated with the terms and thereby can generate in a protein sequence engineering task sequences with desired properties. The Genetic Algorithm starts with a population of 40 sequences consisting of the 102 residues of the N-terminal part of the λ repressor. They contain random bit mutations that do not alter the sequence with the used coding. The Genetic Algorithm generates mutations with changed residues by building new generations using crossover and bit mutations. The Genetic Algorithm selects mutations that establish a balance between the user weighted criteria.

The third application is a protein folding trial of a model protein of a four β strand bundle. The conformation is represented by placing the alpha carbon atoms on a tetrahedral grid. Side chains are not modelled hence there is no distinction between different amino acids. The grid enables four states that the dihedrals of the backbone can adapt which are coded on the strings. The fitness functions penalize backfolds, when atoms occupy the same grid point. They reward substructures that correspond to β strands and a contracted fold, properties of the known native fold of the test protein. In the resulting folds, all of the β strands are formed. The folds obtained with a fitness function considering interactions of strand ends are not unique and the expected optimal fold is not achieved. A fitness function including a term for globularity yielded a unique reproducible fold resembling the expected optimal fold.

5.4 Learning Folding Rules

Judson [11] optimized folding pathways leading to conformations instead of the conformations themselves. They are represented as programs consisting of several rules describing the folding steps. The objective value is the conformational energy of the generated fold. Two-dimensional model polymers are used which consist of atoms linearly connected by harmonic bonds. The energy function contains terms for bond lengths and non bonded interactions. The bond length terms have their minimum at a length of 1. The non bonded terms have a minimum of depth 1 at a distance of 1. The global minimum of the molecule therefore is a hexagonal close-packed structure with each atom having a maximum number of other atoms unit distance away from it.

A folding rule consists of a pair of atom indices. These two atoms are linked with a temporary bond that should bring them together and modify thereby the fold. This is achieved by an conjugate gradient energy minimization and thereafter removing the temporary bond and modifying the atom positions slightly at random. All rules are applied in turn and finally a minimization without a temporary bond is performed leaving the final fold. The approach was tested using a 19-atom polymer, populations of 20 rule sets and each rule set containing 20 atom pairs. The minimum energy attained in the generations finally reaches the known global minimum while remaining at the same energy value over several generations and dropping at the end of the plateaus.

6 CONFORMATIONAL OPTIMIZATION USING GENETIC ALGORITHMS

Protein structure prediction aims to generate the native conformations provided only the amino acid sequence. A common approach is the minimization of the conformational energy calculated using an empirical force field. Genetic Algorithms can be used instead of traditional optimization methods in order to handle vast solution spaces and to circumvent the multiple minima problem.

The driving force of protein folding in nature is the minimization of the free energy. Protein structure prediction by energy minimization in analogy to this changes protein conformations to those with lower energies and tries to reach the final state of this process in nature. It is, therefore, suitable for all possible sequences and depends neither on knowledge of single proteins or protein classes, nor on statistical knowledge obtained from the analysis of known structures. In this way, there is no temptation to anticipate the solution to be found by incorporating its properties into the fitness function. The known structures are only used to verify the solutions obtained with an independent criterion.

The representation of the molecular structure should be as accurate as possible. Strong simplifications as paths on grids and neglection of side chains generate structures that are difficult to compare with data of known structures.

6.1 Representation of Molecular Structure

The conformation of a molecule could be defined by the Cartesian coordinates of its atoms. There are two reasons why they are not suitable as representation in a Genetic Algorithm. First, not all vectors of coordinates represent legal conformations. This matters when randomly initializing the start population. Additionally, the crossover usually will produce unrealistic bond lengths and bond angles at the point where the fragments are joined. Moreover, the fragments may overlap. Second, similar conformations of the same section of different conformations will not exhibit any similarity in the coordinates since any translation or rotation changes them.

Therefore, the dihedral angles of rotatable bonds, termed internal coordinates, are used instead. Figure 5 shows an example. The conformation is completely defined if the bond lengths and bond angles are hold rigid. Any combination of dihedral angles generates a legal conformation irrespective of possible clashes. Furthermore, equal partial conformations in different conformations have the same dihedral angles which supports the formation of partial solutions.

6.2 An Appropriate Selection Mechanism

The properties of empirical force fields impose a difficulty on the regularly used selection mechanism. The selection pressure depends on the distribution of the objective values.

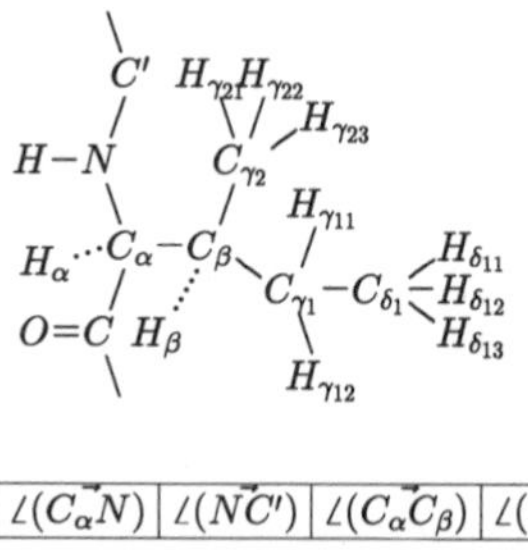

$$\cdots \left| \angle(C\vec{C}_\alpha) \right| \angle(C_\alpha \vec{N}) \left| \angle(N\vec{C}') \right| \angle(C_\alpha \vec{C}_\beta) \left| \angle(C_\beta \vec{C}_{\gamma_1}) \right| \cdots$$

Figure 5. Section of a protein and its coding

Normally, scaling mechanisms are used to ensure that the best member has a certain expected value. These mechanisms use the population average of the objective values. The energy values range over many orders of magnitude and some are extremely large. Figure 6 shows a sorted population with the energy values logarithmically scaled on the y axis. The repulsive part of the van der Waals term rapidly increases the energy if atoms are nearly in contact. A few bad conformations with high energy rise the average to such an extent that the differences of the expected values of the other members disappear.

Therefore, a ranking mechanism, the (μ, λ) Linear Ranking [12], illustrated in figure 7, is used. The expected values for selection are determined by the ranked fitness rather than by relative fitness. The population of size λ is sorted according to the energy values assigning one of the λ ranks to each member. The best member has rank 1 and get the highest expected value. All members get expected values that decrease in equal steps with the rank until the rank μ is reached beyond whom the expected value drops to 0. The remaining worst members receive no offspring during selection. There is a parameter to adjust the selection pressure, illustrated by the arrow in figure 7. The inclination of the dotted line can be adjusted in the range of the full lines, indicating different selection pressures. The greater the inclination of the dotted line the greater the selection pressure.

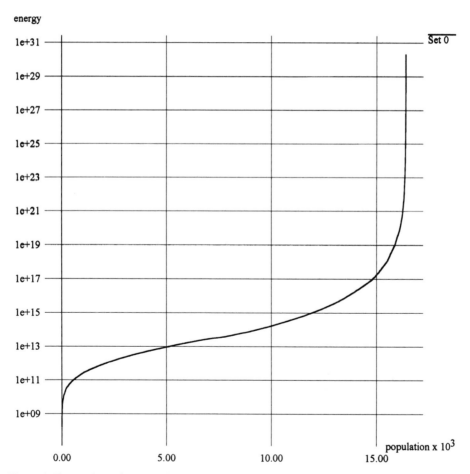

Figure 6. The members of a population are sorted according to their energy values. The members with the highest energy values are only a fraction of the whole population, but their energy values increase dramatically (The energy values are on a logarithmical scale). These members have an impact on the average energy value that disables the use of the standard selection mechanism.

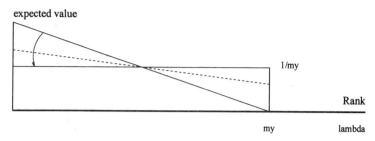

Figure 7. Linear Ranking: selection probabilities according to the rank

6.3 Exploiting Parallelism

Genetic Algorithms are suited for parallelization. The objective values of the solutions are computed completely independent. The objective values are regularly computed one after another. They can be computed all at the same time just as well. If one has access to a parallel computer one should use it since big populations may be required, especially for long proteins, and the computational effort of energy calculation grows quadratically with the number of atoms. Even **Single-Instruction-Multiple-Data-Computers** are suitable for this task since the same computation is performed for all members of the population, while only the rotational angles, respectively, the cartesian coordinates differ from one member to another.

The Genetic Algorithm was parallelized in order to speed up the optimization. The employed massively parallel computer, *MASPAR-MP1* a SIMD Computer with 16384 processors, has a seriel Front End computer. This allows the user to parallelize the most time consuming parts of a program, while running the rest of the program on the Front End. The interface between the seriel and the parallel part is identical with the interface between the core of the Genetic Algorithm and the objective function, the expensive energy calculation. Each of the 16384 processors computes the Cartesian coordinates and the energy of one member. The bit strings are first transmitted from the Front End to the *MASPAR-MP1* then the parallel computation takes place. Finally the resulting energy values are transmitted back to the Front End, where the other parts of the Genetic Algorithm are implemented. The speedup achieved in comparison with a *SUN SPARCstation ELC* is approximately 430. Table 1 shows computation times for different sequence lengths.

6.4 Results

Two force fields are used, the GROMOS force field and the Tripos force field. The sequences of the test proteins are taken from entries in the Brookhaven *Protein Data Bank* (PDB) and from articles that contain representations of generated conformations. The results

Table 1. Speedup on the MASPAR in comparison to *SUN SPARCstation ELC*

No. of residues	*MASPAR* 16 K members	*SPARC* station 1 member	*SPARC* station 16 K-population	speed up
36	2 min 1 sec	3.15 sec	14 h 20 min	426
46	3 min 6 sec	4.9 sec	22 h 18 min	431
54	3 min 53 sec	6.1 sec	27 h 45 min	429
56	5 min 9 sec	8.05 sec	36 h 38 min	427

energy

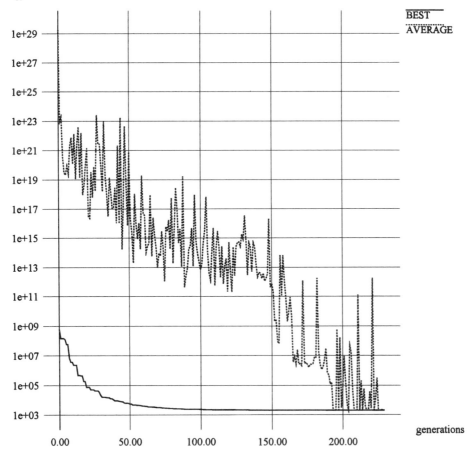

Figure 8. The best and the population average energy values are showed at each generation. The y-axis is logarithmically scaled to visualize leaps over several orders of magnitude.

presented here are from optimizations of Met-Enkephalin using the Tripos force field. The original conformations and the structures energy minimized by the Genetic Algorithm are compared regarding the *R*oot *M*ean *S*quare (RMS) of differences in atom positions and the conformational energy. To compare the energy values, the original conformations first have to be locally optimized using the force field employed by the Genetic Algorithm. Since they were generated by using different force fields, only slight changes in conformation may be necessary to reach the corresponding minimum in the force field used for the Genetic Algorithm optimizations. Figure 8 shows a typical curve of the population best and average energy values illustrating the ongoing progress of a Genetic Algorithm optimization run. The peaks of the average energy that rise after the average energy has reached the level of the best energy are caused by mutations that generate high energy clashes.

A problem called premature convergence appears in Genetic Algorithms when the population converges to a suboptimal solution. Repeated optimizations leading to different solutions indicate that the convergence is achieved too rapidly, due to a too strong selection pressure or a too small population size. The generated structures tend to evolve stretched conformations, maybe since they have on the average less clashes, which causes premature convergence to those structures, before the energy gain of packing the structure can contribute. Figure 9 presents a polyprolin structure obtained with the GROMOS force field. It shows that the attractive part of the van der Waals forces can take effect, when there are less clashes due to missing sidechains.

To avoid premature convergence, the population size has to be adapted to the sequence length of the protein. The results described below are obtained by optimizing Met-Enkephalin, a small pentapeptide, using the Tripos force field. Figure 10 shows the population best energy values of a minimization and figure 11 shows a generated conformation. All optimization runs led to the same minimum when using a population size of 16384, which is the number of processors of the *MASPAR*. The selection pressure was adjusted by setting $\mu = 1.5$ and $\lambda = 16000$ and the crossover rate was 0.6. The conformation was coded by 24 dihedral angles

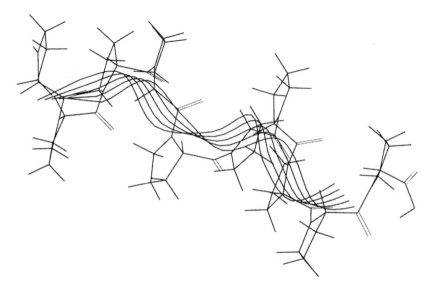

Figure 9. Helical polyprolin structure generated through GROMOS force field minimization

ranging from 0 to 360 degrees, binary coded at a resolution of 12 bits each. The populations converged on the average after 480 generations. Therefore, about 2^{23} points of the search space, containing 2^{288} points, were evaluated, including multiple evaluations of unchanged members.

Table 2 presents the dihedrals of the generated conformation. It can be seen that the ω-dihedrals in the backbone correctly favour the trans configuration of peptide bonds. This structure is refered to as structure No. 0 in table 4 and is compared in table 3 and table 4 to structures given in several articles. Structure No. 1 [13] is the minimum energy structure in the absence of water generated by a Monte Carlo-with-Minimization run using the ECEPP force field [14, 15]. Structure No. 2 [16] is the structure A generated by the *Diffusion Equation Method DEM* using truncated gaussians based on ECEPP. Structure No. 3 [16] is the

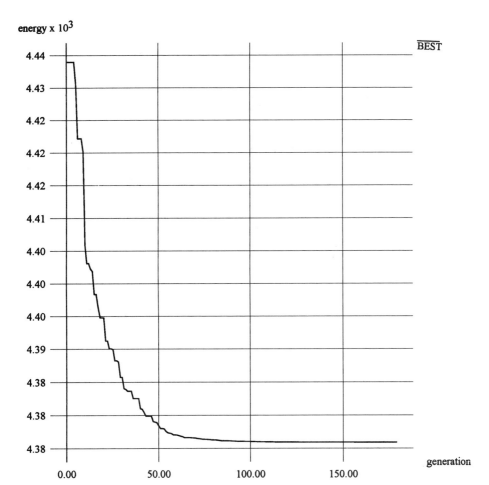

Figure 10. The energy of the best member of each generation. Met-Enkephalin was optimized using the Tripos force field.

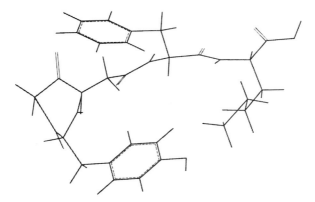

Figure 11. Conformation of Met-Enkephalin generated by a Genetic Algorithm optimization using the Tripos force field.

structure C generated by optimizing structure No. 1 using the *DEM* with the modified ECEPP force field. Structure No. 4 [17] and No.5 [18] are given along with structure No. 1 in the article [13]. The structures are build from the dihedrals denoted in the articles and are compared using the *Sybyl* molecular modelling program. Table 3 presents the *Root Mean Square* of atom positions in Å of these structures and the Genetic Algorithm generated structure. Three *RMS* values are given for the comparison of all atoms, all atoms except hydrogen and only backbone atoms. The backbone of structure No. 0 shows a turn that is most similar to those of structure No. 1 and structure No. 5.

Table 4 compares the energy values regarding the Tripos force field. The structures 1-5 are locally optimized to adapt to the Tripos force field. The energy function used includes terms for bond lengths and bond angles. The Genetic Algorithm optimized the dihedrals and therefore structure No. 0 also is locally optimized to obtain a structure comparable to the optimized structures 1-5. The optimizations are carried out using the *Sybyl* molecular modelling program. The charges are computed by the Pullman method [19] and were also used during the Genetic Algorithm optimizations. A constant dielectricity with a relative

Table 2. Dihedral angles of the conformation generated by the GA

ϕ	ψ	ω	x_1	x_2	x_3	x_4
-116.5	65.0	177.0	-59.1	94.3	34.3	
75.6	-73.1	179.4				
-144.6	48.2	178.2				
-165.1	160.2	-179.6	174.6	87.8		
-133.2	-51.6	178.4	-57.9	75.8	-157.6	62.9

Table 3. *RMS*-Distances to structure no. 0

Structure No.	all atoms	without hydrogens	backbone
1	4.62	4.16	2.41
2	3.64	3.37	2.25
3	4.65	4.22	2.52
4	4.80	4.36	2.50
5	4.39	3.99	1.96

Table 4. Local Minimization using the Tripos force field

Struct.	energy	vdw	coulomb	minimized	vdw	coulomb	RMS
0	17.824	-7.879	9.191	5.588	-12.919	5.505	0.83
1	32.664	9.531	5.202	0.126	-15.491	1.882	0.38
2	36.428	0.578	16.567	18.356	-6.311	13.124	0.6
3	75946.497	75923.914	3.993	13.937	-10.55	11.246	1.05
4	76.024	57.082	2.354	2.104	-10.6	-2.914	0.55
5	83.997	56.885	7.69	4.599	-15.173	7.004	0.47
6	15.033	-8.404	5.169	0.124	-15.469	1.87	0.33

dielectricity of 1 is used. The employed minimization algorithm is the BFGS, a quasi-Newton procedure. Table 4 shows the energy of the original structures, the minimal energy when the algorithm detected convergence and the RMS-deviation from the original structure.

Structure No. 1 reached the minimal energy value while structure No. 0 initially had a lower energy. This could be due to the minimization of bond lengths and bond angles during local optimization or there is a minimum in the search space of dihedrals that is in the vicinity of structure No. 1 and has an energy value below that of structure No. 0. To generate such a structure, Genetic Algorithm optimizations are performed with seeding a fraction of the initial population with the conformation of structure No. 1. In fact such a structure was generated and is refered to as structure No. 6 in table 4. Its energy is below that of structure No. 0 and it converged when locally minimized to the same minimum as structure No. 1. The initial population was seeded with 1024 copies of structure No. 1. A number of 5 and 512 copies were not sufficient to reach the minimum of structure No. 6. This and the fact that all optimizations without initialisation led to structure No. 0 indicates that the search is tracked away from the minimum. Only a sufficient number of initialisation structures has the effect of mainly mutating them by applying crossover.

Objective functions and associated codings with misleading building blocks are called Genetic Algorithm-deceptive problems. Packed structures lie in regions of the search space with high average energies due to the higher probability of clashes. The energy advantage of a minimum in these regions over those in more stretched conformations are too small compared to the high energy conformations surrounding them.

Optimizations with a modified energy function and without seeding the initial population did not find a better minimum than structure No. 0. The energy terms were cut off when the distance fall below a threshold to lower the impact of clashes on the average energies of regions with packed structures. The threshold is defined by multiplying the distance where the energy term is 0 with a certain factor. So, typical cutoff energies were 500 and 170 kcal per pair, according to a factor of 0.55 and 0.6. Not the objective function itself is deceptive but the combination with the coding.

Assuming that the bits of the highest order converge first in each representation of a dihedral angle, optimizations with a different coding were performed. When the bits converge in the order of decreasing value by refining the range of the dihedrals, bringing the bits of the same order together on the bit string could improve the Genetic Algorithm-performance. Therefore the bits were rearranged so that the highest order bits of all dihedrals were followed by the next lower order bits and so on. The optimizations converged in just 330 generations but the generated structures had higher energy values than structure No. 0. This coding is therefore not superior to the coding used so far.

The choice of coding is the point where the sampling strategy of the Genetic Algorithm can be influenced. First a strait forward coding was used. More sophisticated codings have to be evolved to improve Genetic Algorithm-Performance on objective functions of the topology of empirical force field energy functions. It would be best if the Genetic Algorithm optimizes the employed coding along with the energy function. Guidelines for the choice of Genetic Algorithm parameters suited for those topologies and avoiding premature convergence have to be developed. This is useful even if the employed force field does not at once lead to correct predictions since further force fields can be tested given the appropriate Genetic Algorithm parameters and codings to tackle functions of that topology.

Genetic Algorithms are very promising in the field of molecular structure optimization since they are global search methods and can easily be adapted to different sets of parameters describing conformations. They efficiently search vast solution spaces and locate good local optima with a relatively small computational effort.

7 SOFTWARE

Genetic Algorithm novices can get public domain Genetic Algorithm simulators via anonymous ftp to carry out own experiments. User defined objective functions written in C can be easily incorporated. The program GENESIS is available at ftp.aic.nrl.navy.mil and the program

GENEsYs is available at lumpi.informatik.uni-dortmund.de. They each contain a set of objective functions so that tests can be performed at once without supplying an own function. There is a news group comp.ai.genetic and one can subscribe to a Genetic Algorithm mailing list by sending a request to GA-List-Request@aic.nrl.navy.mil.

REFERENCES

1. John Maynard Smith.Evolutionary Genetics.Oxford University Press, 1989.
2. John H. Holland. Adaptation in Natural and Artificial Systems . The University of Michigan Press, 1975.
3. David E. Goldberg. Genetic algorithms in search, optimization, and machine learning . Addison-Wesley Publishing Company, Inc., 1989.
4. Ingo Rechenberg. Evolutionsstrategie . Friedrich Frommann Verlag, 1973.
5. David E. Goldberg. Genetic algorithms in search, optimization, and machine learning , pages 106-120. Addison-Wesley Publishing Company, Inc., 1989.
6. David H. Ackley. An empirical study of bit vector function optimization. In Lawrance Davis, editor, Genetic Algorithms and Simulated Annealing , chapter 13. Morgan Kaufmann Publishers, 1987.
7. David E. Goldberg. Genetic algorithms in search, optimization, and machine learning , pages 204-· 206. Addison-Wesley Publishing Company, Inc., 1989.
8. Marcel J.J. Blommers, Carlos B. Lucasius, Gerrit Kateman, and Robert Kaptein. Conformational analysis of a dinucleotide photodimer with the aid of the genetic algorithm. Biopolymers, Vol. 32, 45-52 , 1992.
9. P.Tuffery, C.Etchebest, S.Hazout, and R.Lavery. A new approach to the rapid determination of protein side chain conformations. Journal of Biomolecular Structure & Dynamics, Volume 8, Issue Number 6 , 1991.
10. Thomas Dandekar and Patrick Argos. Potential of genetic algorithms in protein folding and protein engineering simulations. Protein Engineering vol. 5 no. 7 , pages 637-645, 1992.
11. Richard S. Judson. Teaching polymers to fold. The Journal of Physical Chemistry, Vol. 96, No. 25 , 1992.
12. Thomas Baeck and Frank Hoffmeister. Extended selection mechanisms in genetic algorithms. In Proceedings of the 4th International Conference on Genetic Algorithms , 1991.
13. Akbar Nayeem, Jorge Vila, and Harold A. Scheraga. A comparative study of the simulated-annealing and monte carlo-with-minimization approaches to the minimum-energy structures of polypeptides: [met]-enkephalin. Journal of Computational Chemistry, Vol. 12, No. 5 , pages 594-605, 1991.
14. F.A. Momany, R.F. McGuire, A.W. Burgess, and H.A. Scheraga. Energy parameters in polypeptides. Journal of Physical Chemistry, Vol. 79, No. 22 , 1975.
15. George Nemethy, Marcia S. Pottle, and Harold A. Scheraga. Energy parameters in polypeptides, updating. Journal of Physical Chemistry, Vol. 87 , pages 1883-1887, 1983.
16. Jaroslaw Kostrowicki and Harold A. Scheraga. Application of the diffusion equation method for global optimization to oligopeptides. Journal of Physical Chemistry, Vol. 96 , pages 7442-7449, 1992.
17. J.W. Moskowitz, K.E. Schmidt, S.R. Wilson, and W.Cui. International Journal of Quantum Chemistry: Quantum Chemistry Symposium, Vol. 22, No. 611 , 1988.
18. H.Kawai, T.Kikuchi, and Y.Okamoto. Protein Engineering, Vol. 3, No. 85 , 1989.
19. H.Berthod and A.Pullman. Journal of Chemical Physics, Vol 62 , pages 942-946, 1965.

APPLICATIONS OF ARTIFICIAL NEURAL NETWORKS IN GENOME RESEARCH

M. Reczko and S. Suhai

German Cancer Research Centre
Im Neuenheimer Feld 280
D-69120 Heidelberg, Germany

ABSTRACT

Applications of artificial neural networks in the field of genome research will be reviewed and some more recent developments in neural network research relevant for future applications will be surveyed. The basic definitions for artificial neural networks and neural learning algorithms will be introduced. The applications range from the recognition of translation initiation sites in nucleic acid sequences, the recognition of splice junctions and exons/ introns in mRNA, the detection of uncommon sequences in cDNA, to the prediction of secondary and tertiary structures of proteins from the amino acid sequence, the detection of structural motifs in protein sequences and the classification of protein sequences into functional families. Most applications employ multilayer feedforward networks trained supervised with the backpropagation learning algorithm or self-organising Kohonen maps adapted unsupervised for feature extraction. The most promising developments in neural network research usable in all mentioned applications are new modular network architectures with more problem-tailored connection topologies such as linked receptive fields and recurrent networks with short-term memory capable of modelling any dynamical system using only inductive learning.

INTRODUCTION

The greatest challenges to computer science in the near future emanate from the tackling of highly complex phenomena in nature as occurring in the Human Genome Project [1] or within the projects in the decade of the brain. These phenomena can only be modelled using systems of adequate complexity. In most cases the analytical construction of tractable models is out of the reach of present day theories. An alternative approach is to inductively construct complex, but tractable models using machine learning methods. One of the most successful techniques is the use of artificial neural networks to identify the behaviour or the properties of processes for which complete theories are lacking. The abilities of these systems include those of any general purpose information processing system and are used for tasks as pattern association, feature extraction or the approximation of any static or dynamic system. Given reasonable constraints to the used neural network model, a learning algorithm can develop a system that matches the underlying processes to any desired degree so that the neural model can be used to make predictions that would be very hard to obtain by other modelling methods or

Computational Methods In Genome Research,
Edited by S. Suhai, Plenum Press, New York, 1994

experiment. These qualities explain the wide spread use of neural networks in many scientific applications.

In genome research, a widespread use of neural networks seems to be inevitable facing the exponentially increasing amount of experimental data that has to be interrelated. However, successful applications of inductive learning methods may only be expected if sufficient characteristic data for the adaption and verification of the system is available. The applications described in the following show that there are many cases where interesting properties can be extracted from genomic data using neural networks.

An artificial neural network is an information-processing device consisting of a large number of highly interconnected processing elements called *units*. Figure 1 shows a general version of a unit. Each unit i performs only very simple computations resulting in a single *activation-value* $a_i(t)$. The activation value is transformed into an *output signal* $o_i(t) = f_i(a_i(t))$ using an *output function* f_i. This function is usually the identity function

$$o_i(t) = a_i(t) \tag{1}$$

or a step function $o_i(t) = 1$, if $a_i(t) > \theta_i$, $o_i(t) = 0$ else, with a *threshold* θ_i for each unit. The output signals are propagated on weighted interconnections between the processing elements called *links*. The signals going into one unit are combined into a *net-input* value $net_i(t)$ using a *propagation rule*. This rule normally calculates the weighted sum of the output signals leading into unit i

$$net_i(t) = \Sigma_j w_{ji} o_j(t) \tag{2},$$

where w_{ji} is the weight on the link from unit j to unit i. Different types e of net-input can be propagated with separate weights w_{ij}^e and summated in separate net-inputs $net_i{}^e(t)$. An example is to have different types of links for excitatory and inhibitory connections.

Using an *activation function* F_i the net-input and the current activation value is transformed into a new activation value

$$a_i(t+1) = F_i(a_i(t), net_i(t)) \tag{3}$$

A very common activation function is the sigmoidal Fermi-function

$$F_i(net_i(t)) = \frac{1}{1 + e^{-(net_i(t) + \theta_i)}} \tag{4}$$

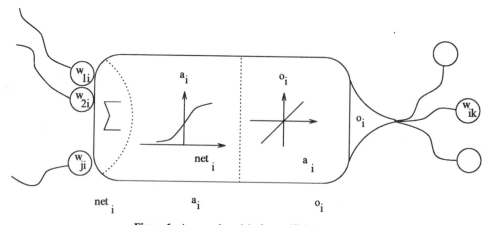

Figure 1. A general model of an artificial neuron.

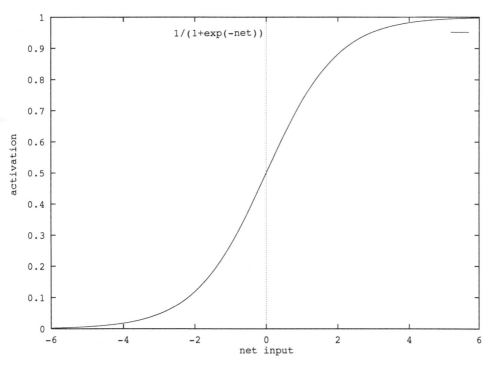

Figure 2. The sigmoidal activation function.

which can be viewed as a model for the dependency between the incoming activation pulses and the resulting spike frequency for a natural neuron (Figure 2).

Interaction with an environment takes place through the exchange of activation values of a dedicated subset of environment units that belong to the categories input, output or both input and output. Units that have no input or output connection are called *hidden* or *internal* units. The operation of a network can be classified into different styles of interaction with the environment during the processing of patterns.

An *auto associator* stores a number of patterns presented at its environment units. During recall, only a part of a stored pattern or a distorted pattern is applied to the environment units and the auto associator adjusts the activations of the environment units according to the stored patterns. With this pattern completion ability, any environment unit may act as an input or output unit.

The most commonly used subclass of an auto associator is the *pattern-* or *hetero-associator* where the environment units are separated into distinct input and output units. The pattern associator performs a mapping from input to output patterns. A subclass of the pattern associator is the *classificator*. Each input pattern is assigned to one out of a fixed number of categories which may be mapped to output activation patterns and back, using a predefined code.

The most important characteristic of a neural network is the topology of the interconnections between all units. A *feedforward* network has no cycles in the connectivity graph defined by the interconnection structure. The units in a feedforward network are arranged into *layers* if all units in a layer have only connections to the units from the previous layer. If there is a cycle in the connectivity graph, the net is called *recurrent* or *feedback* network. Having feedback connections an *update order* for the application of the activation function has to be defined. With *synchronous update* all units in the network calculate their

193

new activation values at the same time. This can be viewed as a discrete difference approximation to an underlying continuous differential equation for each unit. This update method is especially useful for processing a sequence of patterns which are propagated into the network at each update step. An application of this type of network for tertiary structure prediction follows in a later section.

A different type of processing may be performed with *asynchronous update*, where at each time step a unit is chosen randomly to apply its activation function. In such networks a pattern is processed usually by predefining the activation values for some input or output units and applying the activation function several times randomly for the rest of the units. This procedure defines a *relaxation* process for each pattern since the serial update of the units leads to a more and more consistent set of activations. It is used in recurrent *Hopfield networks* [2] and *Boltzmann machines* [3].Networks with saynchronous update are sometimes also called fixpoint-networks opposed to non-fixpoint networks using synchronous update.

The behaviour of a network depends mostly on the interconnection structure which may be adapted by a learning algorithm. There are two paradigms for learning regarding the form of storage of knowledge. The first is *supervised* or *associative learning* were a teacher provides a set of desired activation values for a set of input patterns. The network has to learn to complete the activations on the environment units when a part of these activations is given. The most common case is supervised learning for a pattern associator where a teacher presents pairs of typical input and output patterns. While learning to map input patterns to corresponding output patterns, the network may extract some general features of the input patterns and properties of the mapping that enable the system to predict the correct output activities for input patterns that were not provided by the teacher. This important ability is termed *generalisation*. The *backpropagation* algorithm [4,5] is the most prominent example for a supervised learning algorithm. It can be used to adapt the weights in a feedforward net with hidden units, which is equivalent to the capability of finding a internal representations for processing patterns taking the form of activity patterns on the hidden units. These internal representations are only restricted by the number of hidden units and their connectivity graph but are otherwise not specified by the teacher. As it has been shown by Cybenko [6], these type of networks may perform any linear or nonlinear static mapping from an n_{in}-dimensional space to an n_{out}-dimensional space.

The purpose of the backpropagation is to reduce a mean squared error function that compares the output of the network when an input pattern is applied with a desired target pattern. The set of output activities $o_i{}^P$ are calculated for a set of input activities $i_j{}^P$ using (2) and (4). Having a set of target activities $t_i{}^P$ for each pattern p, the pattern error E^P is defined as

$$E^P = \Sigma_i\,(o_i{}^P - t_i{}^P)^2 \tag{5}$$

and the global error E as

$$E = \frac{1}{N_p}\Sigma_p\,E^p\,, \tag{6}$$

where N_P is the number of patterns used to adapt the weights.

For a fixed input pattern P output activities $o_i{}^P$ depend only on the set of weights w_{jk} and thresholds θ_k in the network. In its simplest form the backpropagation algorithm performs a gradient descent on the error E using

$$\Delta w_{ik} = -\varepsilon\frac{\partial E}{\partial w_{jk}} = -\varepsilon\frac{1}{N_p}\sum_p\frac{\partial E^p}{\partial w_{jk}} \tag{7}$$

194

The backpropagation algorithm determines the partial derivative of the error for each weight. In the *batch-update* version of backpropagation, this is done for each pattern separately and then the results are averaged. A more simple approximation uses the derivative calculated with one pattern for the gradient descent. This is called *online-update* of the weights. For a weight leading into an output unit, the derivative may be calculated using (1), (2), (4) and (6)

$$\frac{\partial E^p}{\partial w_{jk}} = \frac{\partial E^p}{\partial o_k} \frac{\partial o_k}{\partial w_{jk}} = 2(o_k^p - t_k^p)o_j$$

(8)

For a weight leading to a hidden unit k, the derivative may be calculated using the influence that the output activity o_k of this unit has on the error at the output layer. The activity o_k directly influences all units i that have a connection from k to i. The derivative can be calculated by applying the chain rule:

$$\frac{\partial E^p}{\partial w_{jk}} = \sum_i \frac{\partial E^p}{\partial o_i} \frac{\partial o_i}{\partial w_{jk}} = \sum_i \frac{\partial E^p}{\partial o_i} \frac{\partial o_i}{\partial o_k} \frac{\partial o_k}{\partial w_{jk}} = \sum_i \frac{\partial E^p}{\partial o_i} F_i' w_{ki} F_k' o_j$$

(9)

Starting with (8) all partial derivatives may be calculated recursively using (9) by proceeding from the output layer backward through the net. In this way, the term $\frac{\partial E^p}{\partial o_i}$ in (9) is known.

There are many extensions to the basic form of the backpropagation algorithm, which use the gradient information much more efficiently than the method of steepest descent. The most important are methods using conjugate gradients and approximations to second order methods.

The backpropagation algorithm may also be applied to recurrent networks. *Backpropagation through time* is a learning algorithm for recurrent networks updated in discrete time steps (non-fixpoint networks). These networks may contain any number of feedback loops in their connectivity graph. The gradients of the weights in the recurrent network are approximated using a feedforward network with a fixed number of layers (see figure 3). Each layer t contains all activations $o_i(t)$ of the recurrent network at time step t. The highest layer contains the most recent activations at time $t=0$. These activations are calculated synchronously using only the activations at $t=1$ in the layer below. The weight matrices between successive layers are all identical. To calculate an exact gradient for an input pattern sequence of length T, the feedforward network needs $T+1$ layers if an output pattern should be generated after the last pattern of the input sequence. This transformation of a recurrent network into a equivalent feedforward network was first described in [7], p. 145 and the application of backpropagation learning to these networks was introduced in [5]. To avoid deep networks for long sequences, it is possible to use only a fixed number of layers to store the activations back in time. This method of truncated backpropagation through time is described in [8].

A special case of supervised learning is *reinforcement learning* [9] which uses only reward or penalty information given by a teacher or an environment according to the outputs of an adaptive agent. The teacher is not forced to give precise output information for each learning situation as in normal supervised learning. In many situations it is much easier to judge only the outcome of actions or plans. In such cases the learning algorithm has to address the *credit assignment problem* which asks which elements of the system are responsible for a correct or wrong output that may occur with a long temporal delay.

Another situation is *unsupervised* or *self-organised* learning which is performed without the provision of explicit goals by a teacher. The task is to find some statistically important features or regularities in a distribution of input patterns which transform the patterns into a more compact representation [10]. The methods of this field are very frequently used to design a preprocessing step in pattern recognition systems.

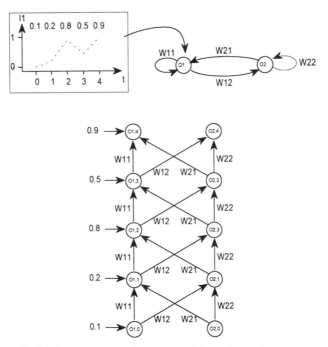

Figure 3. A fully recurrent net with 2 units and the equivalent feedforward net.

TIME DELAY NEURAL NETWORKS

A feedforward architecture especially designed for processing pattern sequences is the *time-delay neural network* (TDNN) developed by Waibel et al. [11]. The usual way of transforming pattern sequences into activity patterns used as an input for feedforward networks is the extraction of a subsequence using a fixed window. This window is shifted over all positions of the sequence and the subsequences are translated into input activities. There are two problems connected with this method. The output information has to be assigned to each subsequence. Therefore the exact relation of the position of the window on the sequence and the desired output information has to be known in advance. This is required since the network will be trained to relate features in the sequences that occur at a fixed position within the input window with output information. If the same features is shifted by one or more positions within the input window, the resulting input patterns are completely different. The network has to learn not only to recognize these features, but must also learn to compensate the segmentation errors.

The second problem arises in those cases when different features are occurring within the subsequence in the input window with different relative distances. This happens very frequently in genomic sequences when one or more elements are inserted or deleted in a sequence. Therefore the network should learn to recognize features in sequences even when they appear with different relative positions. The TDNN architecture addresses these problems by imposing certain restrictions to the network topology and the way weights are updated. All hidden units connected to the input layer may only be connected to a limited number of input units that represent a consecutive patterns of a subsequence in the input window. The hidden units therefore have a *receptive field* that is only sensitive to a part of the input window. The important restriction is that for each receptive field a number of hidden units with the same receptive field has to be assigned so that the receptive field occurs at each position in the input window exactly once. If the input window contains for example 13 positions and a receptive field covers a subsequence of 3 positions, there are 11 hidden units having the same receptive field. Since the corresponding weights in all copies of a receptive field are forced to have the same values, these hidden units have *linked receptive fields*. They may recognize a feature invariant to its relative position within the input window and hence represent one *feature-unit*. During the adaption of the network, the partial derivatives of corresponding weights in linked receptive fields are calculated separately since these hidden units with their receptive field at different positions in the input window get different activations. To adapt a receptive field, the weight update is averaged over all copies of the corresponding weight. This average update is then applied to all copies of that weight. In this way it is ensured that the copies of a receptive field remain identical.

The concept of the linked receptive fields leads to a network where a feature-unit recognizes a subsequence independent of its position within the input window. It is important that the receptive field adapts so that the feature unit recognizes a characteristic subsequence *and* does not recognize other subsequences at *all* other positions in the input window. This leads to a different set of weights compared to a network where a subsequence is trained to be recognized at one position and the weights of this network just are repeated at all positions of a larger input window.

Using the receptive fields of the feature units, the subsequence in the input window is transformed into a sequence of feature unit activations. In its basic form, a TDNN integrates these sequences of feature unit activations using weights from all copies of each feature unit to an output layer or any feedforward topology network. These weights may now express the exact position within the input window at which each sequence feature has to occur. In order to construct more complex structures from the recognized features with some invariance for positional changes of the relevant features, the second layer of hidden units may also have linked receptive fields.

In this way a hierarchical arrangement of layers with linked receptive fields that integrates more and more complex information from the sequence can be constructed. The receptive fields in the higher layers are needed if the relative positions of the features in the input window may vary with respect to each other. There are several successful applications of TDNN's in speech recognition [11] or the recognition of handwritten characters [12].

PREDICTION OF FUNCTIONAL SITES AND FEATURES OF GENOMIC SEQUENCES USING ARTIFICIAL NEURAL NETWORKS

In parallel to the expansion of databases on genomic sequences, there is a steady increase also in the diversity and efficiency of mathematical and informatics methods to predict and analyse functional sites on nucleic acids. For problems related to well defined sites, like those for the recognition of restriction enzymes, the technique of constructing consensus sequence patterns and comparing them automatically with newly sequenced DNA provides a reasonable solution. On the other hand, the establishment of base frequency matrices containing the frequency of each base at each position of a given functional site from a compilation of identified patterns may help to find less well defined signals on the genome. Though the computation of such matrices is a very tedious work, once they have been compiled, they provide much more information than a simple consensus sequence. A further avenue for the development of more sophisticated pattern recognition tools will be opened by the application of neural networks for this purpose.

Beyond the obvious advantages that such tools are much easier to develop than statistical and/or rule based methods and that, once the networks have been trained, they can be easily distributed and employed to search on new targets, we may also expect that neural nets with a complex architecture will make it possible to learn higher order correlations between sequences and functions then it was possible with traditional methods. In fact, there is extensive genetic evidence for context-dependent and/or compensatory mutations in regulatory genome sequences. Their adequate analysis and understanding requires, therefore, methods that accept them in their entirety and make use all of their information, not only position by position, but also cross-correlating every position with every other one. The most important DNA pattern recognition problems, to which several neural net studies were devoted in the past years, include the distinction of ribosome binding sites [13, 14], the analysis of promoter recognition [15-25], and the differentiation between coding and non-coding (exon/intron) regions in eucaryotic genomes [26-42].

IDENTIFICATION OF RIBOSOME BINDING SITES IN E. COLI

The first application of an artificial neural network for a DNA sequence analysis problem was the use of the perceptron algorithm to predict ribosome binding sites on mRNAs by Stormo et al. [13, 14]. In their first paper, they established rules built by including information from the analysis of about 35 bases of mRNA surrounding the initiation codon [13] and collected an mRNA library of 124 identified genes consisting of 78,612 bases. In their subsequent work [14], they have set up a training set of the 124 known genes complemented with 167 'nongenes', i.e., identified false gene beginnings. The perceptron was trained (in three distinct investigations) by the first 101, 71, and 51 bases, respectively, of the training set. It turned out that at least 71 bases were needed for the network to be able to recognize all 124 positive elements in the training set and the network trained with 101 bases was by far the most successful in predicting new genes (with the fewest false positive cases). This is somewhat longer than the sequence expected to be necessary for the ribosome to identify the initiation site and Stromo et al. suspected that their mRNA library might have still contained some hidden positive sites included in the 'nongene' group of the training set and forcing thus the perceptron to look for information in additional bases not relevant to the ribosome itself. Despite of this problem, the network correctly predicted in a test set containing ten genes six proper beginnings and incorrectly identified five false beginnings [14]. As stressed by the authors, the previous method based on rules (that were much more difficult to compile) could only predict five true genes at the expense of twelve false ones.

PREDICTION OF TRANSCRIPTION INITIATION SITES

The proper selectivity of DNA transcription will be insured by the interaction of RNA polymerise with specific DNA sequence patterns, the promoters. The statistical analysis of E. coli promoter sequences [15, 16] led to the identification of two highly conserved hexanucleotides segments, the TATAAT and TTGACA boxes, respectively, situated at the positions -10 and -35 upstream from the starting point of the transcription. The spacer between these boxes varies between 15 and 21 basepairs (bp). Many promoter mutations lie within these regions, with 'up'mutations increasing and 'down' mutations decreasing the homology to the above two consensus patterns. Though it is known that the sequences of both the boxes and of the spacer (and of its length) are essential for the biological function, no clear-cut rules could be established for them, probably due to the huge variability of the individual promoter sequences. On the other hand, sequence regions containing two segments homologous to TATAAT and TTGACA, with the spacing of proper length between them, may occur by chance at several other locations in the genome. Thus, the presence of the consensus pattern does not by itself imply that the region is a promoter.

Biochemical analysis suggests that promoter activity may be related, besides the characteristically high A+T content of the DNA in this region, to a relatively low local melting temperature promoting easier unfolding and to higher order structural features (basepair dependent helical twist, torsion angle, et.). In their discriminant analysis of promoter regions in E. coli, Nakata et al. [17] combined, therefore, the consensus sequence patterns for E. coli promoters, as obtained by the perceptron algorithm, with different combinations of such indices. As a training set, they used 57 true promoters from the Hawley-McClure tabulation

[15] and complemented them with 59 false sequences matching the TTGACA and TATAAT regions by >8 hits, with a spacing of 15-19 bp, and the regions were located >50 bp downstream from the true promoter region. They applied another set of 33 true and 43 false sequences for prediction.

The perceptron achieved 100% discrimination for the training set of sequences, but it gave only 67.1% for sequences outside of the training set. In contrast, the base composition, the melting temperature and the three-dimensional structure of the double-helix were relatively unaffected in their power of discrimination whether sequences to be discriminated were within or outside of the training set. This experience motivated Nakata et al. to combine the perceptron value, that turned out to be very sensitive to the choice of the training set, with the more robust other variables computed. In fact, different combinations of this properties led to an increased discrimination power, up to 75%. Application of the perceptron discriminant functions, as determined by the E. coli training set, for the prediction of promoters in other organisms, provided a success rate of 60% in other bacteria and 49.7% in yeast. Combination of the perceptron values with other variables improved the situation again to 62.7% and 52.8%, respectively. Apparently, the training sets turned out to be organism-specific.

Lukashin et al. tackled the same problem by using a specific neural architecture consisting of separated blocks of three-layer networks that were trained independently. They presented to each of these blocks only promoter segments of limited length belonging to certain regions of the promoters (e.g., -10, -35, etc.). The motivation for such a subdivision of the problem was the observation of these authors that, assuming an average promoter length of 70 bp, the information content of the presently available experimental data is insufficient for the problem to be treated completely with consideration of all possible correlations in the sequence over the entire length of the promoter. (Through the above decoupling procedure, of course, all such correlations will be lost.) On the other hand, experimental data on the efficiency of mutant and artificial promoters suggest that correlations within the -10 and -35 boxes, the spacer, and the downstream region are more important than cross correlations between these regions [18]. On the other hand, the intentional restriction of the network capacity insures its capability of generalization in view of the size of the actually available training set.

For training, Lukashin et al. selected 25 'strong' promoters from the Harley-Reynolds collection [16] and complemented them with 250 'nonpromoter' random sequences of the same length, in which the four nucleotides occurred with equal frequencies but the sequences fulfilled the boundary condition of reducing a specific error function [18]. Each neural net block had 12 neurons in the input layer and four neurons in the hidden layer. Their single output units were coupled. After training, the testing set was chosen as the 222 sequences with known transcription initiation sites [16]. Pairs of hexanucleotides 15 to 21 bp apart were distinguished near to positions -10 and -35, respectively, and each of the 222 sequences was represented as a set of such pairs during testing. 'Nonpromoters' were represented again as 2220 pairs of random hexanucleotides with a uniform base frequency.

In analyzing the performance of their network, Lukashin et al. stress that a clear distinction has to be made between the successful identification of true promoters and the error leading to false ones. They obtain 94 to 98% of the correct promoters for the testing, i.e., the probability of a pair of random hexanucleotides being identified as a promoter varies in between 2 and 6%. In the case of the pBR322 plasmid, for instance, which has a length of 8726 nucleotides for both strands, the algorithm correctly identified all six promoters but it yielded, depending on certain network parameters, 9 to 180 extra sites. This error is comparable, however, to that of the most advanced homology score methods [19, 20].

Demeler and Zhou employed a simple backpropagating network for E. coli promoter prediction [21] but they paid special attention to the design of the training and test sets (especially to the ratio of promoter and random sequences), to different forms of input representation, to the number of hidden units, and to the error level to be achieved during training. They selected 80 bacterial and phage promoters of the Hawley-McClure compilation

[15] as the training database and used the remaining 30 plasmid and transposon promoter sequences and promoters generated by mutation as a test set to evaluate the method. The promoters were arranged into three independent training and corresponding test sets comprising of 20 bases centred around the -10 region (set A), 20 bases centred around the -35 region (set B), and 44 bases including both regions together without the spacer. The promoter sequences were complemented again with random ones in the training sets using ratios of 1:1 to 1:20. The prediction accuracy was measured as the fraction of sequences in the test set predicted correct. Predictions were performed independently on promoter sets and random sets, respectively.

Demeler and Zhou found that the number of hidden units does not have a significant effect on the prediction accuracy. Important is, on the other hand, the 4 bit-representation of the bases in the input layer (preferred over 2 bits). A low network error level improved the promoter prediction accuracy, while larger random-to-promoter (r:p) ratios improved the network's ability to filter out false positives. The optimal r:p value was found between 5:1 and 10:1. Interestingly, significant differences have been observed between predictions of the -10 and -35 regions, respectively, presumably due to a larger correlation between the promoter sequences in the -10 set. This suggests that the -35 region is less conserved than the -10 region in E. coli promoters (an average difference of 23% was observed with set A giving the better result). Although the prediction accuracy of training set B was substantially lower (68.4%) than that of A (90%), it was still above 50% (expected if no classification could be made at all). This suggests that the information content of B could add to A. In fact, combination of A and B as in C provided a prediction accuracy of 98.4%. As for previous methods, the price for this accuracy is again substantial overprediction.

O'Neill reported in two recent papers [22, 23] on the training of three-layer backpropagation networks to recognize promoter sequences of each of the three major spacing classes found in E. coli (with 16, 17, and 18 bp, respectively). For the 17 bp spacing class [22], the network was presented with 39 promoter sequences and their derivatives as positive input, and with random sequences and sequences containing 2 promoter-down point mutations as negative inputs. He tested more than 25 distinct network configurations, including networks with two hidden layers containing 3 to 60 neurons per layer. Of the architectures explored, the network employing 58x4 input neurons, one hidden layer of 15 interneurons, and a single output neuron was found to be the most effective. For the input presentation, the 4-bit per nucleotide coding was found again superior as compared with the 2-bit code. O'Neill's input consisted of 5148 58-base sequences drawn from the 39 promoters of the 17-base spacing group, 4000 random sequences with 60% A+T content, and about 100 examples of the P22 ant promoter representing all pairwise permutations or double promoter-down mutations from the list of strong promoter-down mutations reported by Youderian et al.[24]. The two smaller groups were duplicated until their size equalled the size of the promoter group, resulting in an input of over 15,000 sequences.

The most important general conclusions of this study are: (i) Backpropagating networks can learn the training set perfectly, can generalize to a high level, and maintain the false positive level in the range of 0.1%. (ii) The major limitation of their success lies not in their inherent power but in the fact that the training task is slightly at odds with what is actually being sought, the ability to generalize. A network trained on an input of 32,000 distinct sequences, e.g., learned after 8,000,000 cycles to correctly classify every sequence, but it was not particularly good on new inputs. Thus the training set has to be designed to be as difficult as possible while remaining faithful to the true and false input categories. (iii) The network specifically trained for the 17-base spacing class improved the results of the best rule-based methods, that reach 70% correct predictions for this particular problem [25], by about 10% as demonstrated on pBR322 and on the lambda genome.

In a subsequent work [23], O'Neill specifically trained further backpropagating networks (using the same architecture as before) to recognize promoter sequences of the other two major spacing classes in E. coli (with 16 and 18 bp, respectively). The network resulting for the 16 bp class captured between 78 and 100% of previously unseen promoters in different tests, while maintaining the level of positive identifications at a fraction of 1%. The network for 18 bp promoters identified 79% of the test cases. A tandem poll of networks for all three spacing classes produced a cumulative false positive level of less than 0.5%. The weight matrices used by the networks during their classification could be analysed to determine the relative weight assigned to the occurrence of a given base at a given position within the promoter. In this way, an approximate description of the network's definition of the promoter could be obtained.

IDENTIFICATION OF CODING REGIONS IN GENOMIC DNA

Recent advances in large scale genomic DNA sequencing technology make it possible to clone and sequence a huge number of genes without their tedious biochemical and genetic characterization in advance. Since, on the other hand, the genes in higher eukaryotes may span tens or hundreds of kilobases with the protein coding regions accounting for only a few percent of the total sequence, this technique requires efficient algorithms for the identification of the genes themselves within large regions of uncharacterized DNA. In the cell nucleus, the non-coding regions (introns) will be excised from the pre-mRNA and the mature mRNA will be formed by concatenation of the coding regions (exons). The splicing process recognizes sequences at the exon/intron and intron/exon borders, donor and acceptor sites, respectively. Though the splicing sites can be characterized by consensus sequences at the 5′ and 3′ termini of the intron, their pattern is not sufficient to identify the actual splicing sites with the required accuracy. The 5′consensus was found to be (C/A)AG-GT(G/A)AGT [26], whereas the 3′ site and the branch point are weakly defined and seem to be specified by an intervening polypirimidine tract characterized by the strict absence of AG dinucleotides [27].

From the mathematical point of view, the general problem of classifying fragments of DNA sequence according to their function has been addressed by several methods, primarily either based on the compilation of tables of codon usage [28-30], on detection of local non-randomness in the sequence [31, 32], or observing the specificities in the nucleotide distribution near start codons and splicing sites [33, 34]. These methods reach, however, usually only a predictive accuracy of 70-80%. Stormo et al. [35] applied first the perceptron algorithm to this problem and though they obtained good training results, the reliability of their network was too low for sequences outside of the training set. Kudo et al. constructed a finite state automaton as a recognizer of 5′splice sites but arrived only at 50-55% prediction accuracy on test genes [36]. Nakata et al. [37] combined the output of two two-layer perceptions (trained to recognize exon/intron and intron/exon boundaries, respectively) with physicochemical and base composition properties in the framework of a discriminant analysis. The perceptron was more accurate than Fickett′s function (a combined measure of base composition and periodicity [31]) for predicting the start of coding regions (84% vs. 74%), the end of coding regions (78% vs. 61%), the exon/intron boundary (91% vs. 66%), and the intron/exon boundary (82% vs. 65%).

Brunak et al. [38, 39] constructed the first back-propagating neural net to realise a joint scheme where prediction of transition regions between exons and introns regulates a cut-off level for splice-site assignment. The combination of the networks is shown in Figure 4. Their input window for the DNA sequences (S_w) varied between 11 and 401, and the number of hidden units (S_h) in the range from 0 to 200. From the total data set of 95 genes the first 65 (part I, 385,488 bp) were used for training and the rest (part II: 30 genes, 190,987 bp) for tests. The 95 genes contained 449 donor and acceptor sites, with 331 in part I and 118 in part II. Different training runs were performed on complete sequences and on short (101bp) segments around the splicing sites, respectively. For this second case, the full training sequence consisted of 33,431 bp, and the test set of 11,918 bp. The relation between correctly predicted

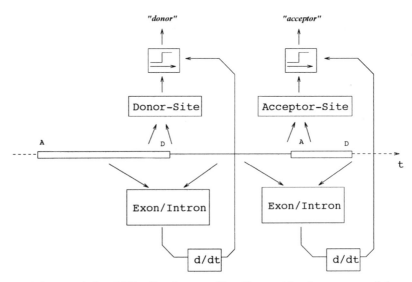

Figure 4. A modular network for mRNA splice site recognition. The transitions between exons/introns modulate the threshold for accepting acceptor or donor sites.

positives and negatives as well as false positives and negatives can be best characterized by a correlation function, C [38], taking the value +1 for perfect and -1 for truly imperfect predictions, respectively.

For the prediction of a nucleotide function as being splicing donor or not, the network trained on the full sequences with $S_w = 15$, $S_h = 40$ arrived at C = 0.9244. It correctly classified 111 of the 118 donor sites (94.1%), and 11,789 of the 11,800 non-donor sites (99.9%). Trained on 101 bp segments, it reached C = 0.41 when tested on part II. It again correctly classified 118 donor sites, but incorrectly classified 536 of the 190,987 non-donor sites as being donor sites. The optimal network architecture for predicting acceptor sites embedded in sequence segments had the parameters $S_w = 41$ and $S_h = 20$. Despite the increased input capacity (with $S_w = 40$ the network precisely covers the polypyrimidine tract of the introns), the predictive performance is lower in this case: 100 of the 118 acceptor sites are correct (87.4%) and 11,778 of the 11,800 non-acceptor sites (99.8%), the correlation coefficient being 0.83.

To test the network's capability of classifying coding and non-coding regions, it was trained with input window sizes of 101 to 401 and with 200 hidden units. At the optimal value of $S_w = 301$, the network correctly identified 69.8% of the translated exons. The combination of the previous (local) information on donor site prediction with this global knowledge of the network on coding/non-coding, substantially reduced the number of false positive donor assignments (by a factor of up to 30). It has to be stressed, however, that both the separate network and the joint assignment method performed considerably better than the weight matrix method of Staden. For a 90% detection of true donor sites, Staden's method assigns 0.412% false positives, whereas the separate network and the joint method assign 0.238% and 0.054%, respectively. For 95% detection of true donor sites, the corresponding values are 0.703%, 0.289%, and 0.094%. When the local information for acceptor sites will be joined with the coding/non-coding network, the number of false positive acceptor site assignments will be reduced again by a factor of up to 40. Comparison with the weight matrix method proves again its superiority: For 90% detection of true acceptor sites, Staden's method assigns 1.3% false positives, whereas the separate and the joint networks assign 0.95% and 0.162%, respectively. For 95% , the corresponding values are 1.986%, 1.606%, and 0.397%.

Uberbacher and Mural [40] designed a 'coding recognition module' consisting of a back-propagation neural network whose seven input nodes received their information from seven specific 'sensors'. They applied two hidden layers (with 14 and 5 nodes, respectively), and 1 output node. The sensor algorithms evaluated different properties of the input DNA sequence that could be relevant in relation with its coding capacity. They included the frame bias matrix (the nonrandom frequency with which each of the four basis occupies each of the three positions within codons), the coding propensity after Fickett [31], a term characterizing the nonrandomness of the dinucleotide distribution, and different preferences in the frequencies of 6-letter words, including their repetitiveness. Input to the network for both training and test sets consisted of a vector containing the values of the seven sensor algorithms calculated for positions at intervals of 10 bases along the sequences of interest. A further specificity of the Uberbacher-Mural procedure is that their network has not been designed to find exons *per se*, but rather, exons, or portions of exons, that encode protein. The training set consisted of 200,000 bases sequence data from GenBank. To evaluate the performance of the network, 19 human genes were examined, none of which were included in the training set. The network located 71 of the 79 coding exons that were longer than 100 bp. It apparently has difficulties with shorter exons (<100 bp) for which it identified only 14 of 30. In terms of sequence positions tested, of the 17,576 locations 16,592 (96%) were correctly assigned as coming from either coding or noncoding DNA. Of the 1113 test points classified as coding, 1029 (92%) were correctly assigned. With regard to the sensitivity of the network to DNA sequencing errors, the experience of these authors suggests that up to 1% insertion/deletion errors can be tolerated.

A significant increase in the prediction accuracy could be achieved for the exon/intron localization problem by Farber et al. [41] who made use of the 'mutual information' (a correlation measure) of spatially separated codons in exons and in introns. Significant mutual information was found in exons, but not in introns, betwen adjacent codons, suggesting (in accordance with the studies of Uberbacher and Mural [40]) that dicodon frequencies of adjacent codons are important for intron/exon discrimination. A slight further improvement was possible by the inclusion of codon and dicodon statistics in all six reading frames (including the complement strand). These authors compare in detail the performance of methods based on Bayesian statistics and emphasize the disadvantages using independent codon frequencies in each position. An interesting further development of the neural network Ansatz can be found in the paper of Snyder and Stormo [42] who combine information from several properties of genomic DNA (codon usage, local compositional complexity, 6-tuple frequency, the characteristic length distributions of exons and introns, splice signals, et.), as extracted by the neural network, with the dynamic programming (DP) algorithm to enforce the constrains that exons and introns must be adjacent and non-overlapping. By DP, they find the highest scoring combination of exons and introns subject to these constrains. Weights for the various classification procedures were determined by training a simple feed-forward network. In a pilot study, the system has been trained on a set of 56 human gene fragments containing 150 exons in a total of 158,691 bps of genomic sequence. When tested against the training data, it identified 75% of the exons and correctly predicted 86% of coding nucleotides as coding while 13% of non-exons as false positives. Because of the simplicity of the network architecture, the generalization performance was found to be nearly as good as with the training set.

APPLICATIONS FOR PROTEIN SECONDARY STRUCTURE PREDICTION

One of the most prominent applications of neural networks trained with the backpropagation algorithm is the prediction of secondary structures using only a local context of the amino acid sequences done by Qian and Sejnowski [43]. All amino acids should be classified belonging to either α-helix, β-sheet or coil-structures. A cascade of two feedforward networks was used. The first network used as input a window of 13 amino acids coded with 21 input units. The 21st unit is employed to fill up the input window at the beginning and end of the sequences. The hidden layer contains 40 units and the output layer 3 units representing the types of secondary structure. Since there is a high correlation between successive positions in a secondary structure assignment, a second network can further improve prediction accuracy. A sequence of 13 contiguous predictions of the first network constitutes the input to the second network with a hidden layer of 40 units and 3 final output units. This cascade has been trained with 91 proteins and tested with 15 proteins with very low sequence homology to the proteins in the training set. Their best result was 64.3% residues predicted correctly. When using a neural network for postprocessing the results of a primary network, great care has to taken to ensure that the second network does not detect irrelevant fluctuations in the output of the first network. The validation with the test-set should also be done in such a manner that the network will not be optimized indirectly for the best result on the test set unless this set is at least as large as the training set.

There have been several similar efforts for using feedforward networks in this area [44, 45] with comparable results. The secondary structure prediction networks developed by Bohr et al. [46] were also used to define a homology measure between sequences. A network for predicting secondary structure or the amino acids based on the local sequence context of one protein is tested for predicting the same properties in another protein. The larger the homology between these two proteins, the better the prediction accuracy.

A major improvement was achieved by Rost et al. [47] using a primary sequence representation that contains much more evolutionary information than the bare sequence. In the HSSP database [48] a distribution is given for each residue which states by what other residue each residue may be exchanged without affecting tertiary structure. This sequence-profile information is found by multiple alignments with all known sequences and defines highly conserved sequence elements that might be essential for defining a structure. It also effectively enlarges the training set size by at least one order of magnitude since many alternative sequences defining one structure are introduced. In the application, the normalized distribution is transferred to a set of 20 input units for each position in the input window. An unknown sequence first has to be aligned against the sequence database to provide the sequence profile information. The choice of this representation alone raises the prediction accuracy to about 69% for three classes. A carefully balanced composition of the training together with the training of a committee of 12 networks that cast a majority vote for each residue led to an accuracy of 71%.

APPLICATIONS FOR THE PREDICTION OF PHYSICOCHEMICAL PROPERTIES OF PROTEINS

An important constraint for determining the tertiary structure of proteins is the knowledge of solvent accessibility for each part of the amino acid chain. Holbrook et al. [49] have designed neural networks for predicting the degree of surface exposure for each residue into either two or three categories. A two layer network without hidden units used a sequence context of 9 residues to predict the central amino acid being either buried or exposed. This prediction has an accuracy of 72% which indicates that the flanking residues have only a small effect on the surface exposure since a prediction based only on the type of the amino acid under consideration achieves an accuracy of 70% for two categories. A three-state prediction using a three layer network with 7 residues as input, 10 hidden units and 3 output units has an accuracy of 54.2%. A possible extension to this system would be the use of a recurrent network that is able to store more long-range information about the presence and relative distance of other hydrophobic parts of the protein that is essential to improve the predictions that use only local information.

A typical long-range interaction in proteins is the formation of disulfide bonds between two cysteines. Despite its non-local character, Muskal et al. [50] were able to predict the presence of a disulfide bond in 81% of the cases correctly using only a local sequence context of 7 residues in each direction apart from a cysteine. This indicates that the local environment around the cysteine is of significant importance in determining advantageous conditions for the formation of a disulfide bond. Since a two layer network was used to learn the prediction, the weights of the network could be interpreted directly in terms of disulfide-bonding propensity. This revealed interesting positional relationships for several amino acids.

Hirst and Sternberg compared the performance of a two layer perceptron with a statistical method for the prediction of ATP/GTP-binding motifs in protein sequences [51]. Using a consensus pattern for this motif 349 sequences were extracted from a protein sequence database. Of these 193 were determined as correct ATP binding motifs and 156 were false positives. Segments of 17 residues from these sequences were used for training and testing the perceptron net. The average performance of the network when trained with all but one sequence and tested on the remaining sequence was a 82% correct prediction of the binding sequences, a 74% correct prediction of non-binding sequences and a 78% correct prediction of all sequences. A statistical method based on a comparison of the alignments of the test sequence with all remaining sequences using a residue similarity matrix (Dayhoff et al. [52]) resulted in a 79% correct prediction of the binding sequences, a 81% correct prediction of non-binding sequences and a 80% correct prediction of all sequences. The statistical method was marginally better than the perceptron net. A network with hidden units was expected to perform at least as well as the statistical method but was not analyzed in this study because of the limited number of training patterns available.

Wade et al. used a three layer feedforward network to predict water binding sites on proteins [53]. The input to the network consisted of a window of 17 residues centered around the residue for which predictions are being made coded using 7 physicochemical properties together with the secondary structure information for the central residue. Two types of predictions were investigated. The first is to determine whether a residue has a water ligand or not using one output unit and the second predicts for each atom whether a water molecule makes close contact with it. For the second type of prediction, two units indicated for each atom in each residue, except C and H atoms, whether a water molecule is within a distance of 3.5 Å of the atom and whether a protein atom which is not in the same or adjacent residue is close to that atom. The network was trained using 40 nonhomologous proteins with a resolution < 2.0 Å and 6913 residues. The test set contained 2224 residues from 9 proteins. The percentage of correctly assigned sites in the training set was 89% and 64% in the test set. The proportion of the true water binding sites predicted as such was about 40% for the test proteins. This compared favourably with other methods to predict water binding sites using tertiary structure information with a corresponding performance of about 40%. The performance of this network might be increased using more accurate water binding site information in the training set and a prediction method with short-term memory in order to consider the cooperative effect when binding several water molecules that influence each other.

APPLICATIONS FOR TERTIARY STRUCTURE PREDICTION

An approach to tertiary structure prediction employs associative memories similar to Hopfield nets to store a finite set of three-dimensional protein structures. Friedrichs et al. [54] construct a *Hamiltonian* function which explicitly contains distance information and information coupled with the type of the amino acids of all proteins to be stored in the memory and the sequence information of a protein that has to be recognized. This Hamiltonian may be interpreted as the weights in a recurrent Hopfield network. Using simulated annealing the coordinates of the C_α-atoms of the protein to be recognized are modified to minimize the value of the Hamiltonian function. To extend the generalisation capabilities for the recognition of proteins with a lower degree of homology an explicit term will be introduced that punishes point mutations in accordance with the probability that this mutation will change the tertiary structure. These probabilities were taken from the investigation by Dayhoff et al. [52].

Versions of the stored proteins containing a limited number of insertions and deletions were generated explicitly and stored in the memory in order to achieve invariance for insertions and deletions in the protein to be recognized. Since these variants are no longer uncorrelated with respect to each other, the capacity of the associative memory is reduced significantly. An experiment to generalize from two stored proteins (*Rhodospirrillum rubrum* cytochrome c3 (2C2C) and *Desulfovibrio desulfuricans* cytochrome c3 (1CY3)) to the 111-residue rice cytochrome *c* (1CCR) with a sequence homology of 42% from 2C2C and 14% from 1CY3 to 1CCR, resulted in a structure with a R.M.S. deviation of 5.3 Å to the native structure (4.0 Å not including the first 13 residues). Although additional information relevant to the folding process may be integrated into the Hamiltonian function in a straightforward manner, the question remains whether the capacity of this type of associative memory can be extended to include all prototypes that are required to model the folding of any protein.

A completely different approach to protein tertiary structure prediction was developed by Bohr et al. [55]. The tertiary structure information is predicted in the form of binary distance constraints for the Cα-atoms in the protein backbone using the primary sequence as input to a feedforward network. The distance constraints are then used to generate a conformation of the protein backbone that minimizes the number of violated constraints. This is done using a steepest descent minimization with a first order approximation to the gradient of the function evaluating the violated constraints. The input to the network was a window of 61 residues shifted over the primary sequence. The network predicts for each distance between the Cα-atom of the residue in the center of the input window and the 30 Cα-atoms of the residues preceeding this residue, whether this distance is larger than 8 Å or not. This corresponds to the prediction of a band along the main diagonal of the binary distance matrix for a protein. This is a square matrix in which the rows and colums are ordered as the primary sequence and each cell contains the binary information whether the distance between the two Cα-atoms of the corresponding residues is smaller than some threshold or not. To give some additional information, the feedforward net should also predict the secondary structure for the residue in the center of the input window.

The network was trained with 13 proteases and when tested on the training set capable to predict the distance matrix output to 99.9% correct and the secondary structure assignment up to 100% correct. A test with rat trypsin (1TRM), not contained in the training set gave a distance matrix band with nearly correct secondary structure features and an accuracy of 96.6%. Significant homologous trypsin and subtilysin were members of the training set, however. The steepest descent optimization was starting with the backbone conformation of a protein 74% homologous to rat trypsin. The final structure had a RMS deviation of 3Å.

APPLICATIONS FOR PROTEIN FAMILY CLASSIFICATION

A successful application [56] of an unsupervised learning algorithm uses the self-organising Kohonen feature map for clustering proteins into families based on their sequence similarity. The sequence of a protein is represented by the normalised frequencies of bipeptides. The input to network is a 20 x 20 matrix containing the number of occurrencies of a particular ordered pair of amino acids divided by the sequence length. The Kohonen algorithm used a two-dimensional map of 2 x 2, 4 x 4, 7 x 7 , 8 x 8, and 15 x 15 units. In a first experiment 10 proteins belonging to 3 different families were separated optimally using the 2x2 map, since the number of neurons is nearly the same as the number of classes. In a 4 x 4 map, the proteins families are arranged into non-overlapping regions with one or more winning neurons. This map was used to test the ability of detecting homologies between unseen sequences and the proteins in the training set. The training sequences were mutated at 1% to 100% of the sequence positions and classified by the net. These sequences were still recognized when 21.5% ± 7% of the positions were mutated. The network was also tested to classify initial, middle and final sequence fragments of the training proteins. It was observed that the correct winner could be found with only 7.5% ± 3% of the original sequence. Using a 7 x 7 feature map [57], a set of 477 proteins could be correctly clustered into 13 different families in 96.7% of the cases. If only one family of proteins is presented to the network, the arrangement of the winning neurons reflects phylogenetic relations of the proteins. A set of 76 cytochrome c sequences was classified into four larger groups: plants, fungi, protozoa and vertebrates. Sequences with their corresponding winning neurons close to each other also have a high degree of sequence homology. Since the activities in the feature map may be calculated very quickly, this system is a powerful tool for searching homologies in large protein databases. Another application could be the selection of an appropriate prediction system that has been optimized separately for each protein.

SUMMARY

There are many successful applications of neural networks for genome research with performances superior to comparable statistical methods. It has to be emphasized that the theory for simple neural network architectures such as the perceptron is very well developed and has shown that these models can be viewed as variants of statistical classifiers or approximators. The most severe problem of all these supervised learning methods is overfitting which is not well respected in some applications where the real performance on a test set is not emphasized enough. More advanced networks such as recurrent networks or complex ensembles of different neural network modules have no comparable statistical model as couterparts and are thus tools to solve new classes of problems that could not be tackled successfully yet. With the increase of neurobiological knowledge this situation will shift much more in the direction where more and more complex and realistic models may only of parts of biological neural networks may be build and applied to difficult pattern processing tasks and these models may only be coarsely analyzed using complex system theory. Yet these new forms of information processing most probably will shown to be of enormous use as a first step in solving extremely hard problems in molecular biology.

REFERENCES

1. J. Watson, Science 248:44-49 (1990).
2. J.J. Hopfield, Proc. of the Nat. Acad. of Sci. USA, 79:2554-2558 (1982).
3. G.E. Hinton and T.J. Sejnowski, Learning and Relearning in Boltzmann Machines, in Parallel Distributed Processing1, eds.: D.E. Rumelhart and J.L. McClelland, MIT Press, Cambridge, Mass. (1986).
4. P. Werbos, Ph.D. Thesis, Harvard University (1974).
5. D. E. Rumelhart, G.E. Hinton, R.J. Williams, Learning internal representations by back-propagating errors. In Parallel Distributed Processing, eds.: D.E. Rumelhart ad J.L. McClelland. MIT Press, Cambridge (1986).
6. G. Cybenko, Continuous valued neural networks with two hidden layers are sufficient, Technical report, Departement of computer science, Tufts university, Medford, MA.
7. M. Minsky and S. Papert, Perceptrons: An Introduction to Computational Geometry, The MIT Press, Cambridge, Massachusetts (1969).
8. D. Zipser, Subgrouping Reduces Compexity and Speeds Up Learning in Recurrent Networks, in Advances in Neural Information Processing systems 2, ed. D.S. Touretzky, Morgan Kaufmann, San Mateo, California, 638-641 (1990).
9. R. S. Sutton, Machine Learning 3:9-44 (1988).
10. T. Kohonen, Self-organization and associative memory, 2nd edn. Springer Berlin (1982).
11. A.H. Waibel, T. Hanazawa, G. Hinton, K. Shikano, K. J. Lang, IEEE Transactions on Acoustics, Speech, and Signal Processing, 37:328-339 (1989).
12. I. Guyon, P. Albrecht, Y. Le Cun, J. Denker, W. Hubbard,Proceedings of the International Neural Networks Conference, Paris, France, p.42-45 (1990).

13. G.D. Stormo, T.D. Schneider, and L. M. Gold, Nucl. Acids Res., 10, 2971 (1982).
14. G.D. Stormo, T.D. Schneider, L. M. Gold, and A. Ehrenfeucht, Nucl. Acids Res., 10, 2997 (1982).
15. D.K. Hawley and W.R. McClure, Nucl. Acids Res., 11, 2237 (1983).
16. C.D. Harley and R. Reynolds, Nucl. Acids Res., 15, 2343 (1987).
17. K. Nakata, M. Kanehisa, and J.V. Maizel, Jr., CABIOS 4, 367 (1988).
18. A.V. Lukashin, V.V. Anshelevich, B.R. Amirikyan, A.I. Gragerov, and M.D. Frank-Kamenetskii,J. of Biomol. Struct. & Dyn., 6, 1123 (1989).
19. M.E. Mulligan, D.K. Hawley, R. Entriken, and W.R. McClure, Nucl. Acids Res., 12, 789 (1984)
20. M.E. Mulligan and W.R. McClure, Nucl. Acids Res. 14, 109 (1986).
21. B. Demeler and G. Zhou, Nucl. Acids Res., 19, 1593 (1991).
22. M.C. O'Neill, Nucl. Acids Res. 19, 313 (1991).
23. M.C. O'Neill, Nucl. Acids Res. 20,3471 (1992).
24. P. Youderian, S. Bouvier, and M.M. Susskind, Cell, 10, 843 (1982).
25. M.C. O'Neill, J. Mol. Biol., 207, 301 (1989).
26. S.M. Mount, I. Petersen, M. Hinterberger, A. Karmas, and J.A. Steitz, Cell, 33,509 (1983).
27. C.W.J. Smith, E.B. Porro, J.G. Patton, and B. Nadal-Ginard, Nature, 342, 243 (1989).
28. R. Staden and A.D. McLachlan, Nucl. Acids Res., 10, 141 (1982).
29. R. Staden, Nucl. Acids Res., 12, 505, 551 (1984).
30. M. Gribskov, J. Devereux, and R.R. Burgess, Nucl. Acids Res., 12, 539 (1984).
31. J. Fickett, Nucl. Acids Res., 10, 5303 (1982).
32. T.F. Smith, M.S. Waterman, and J.R. Sadler, Nucl. Acids Res., 11, 2205 (1983).
33. M.B. Shapiro and P. Senepathy, Nucl. Acids Res., 15, 7155 (1987).
34. M.S. Gelfand, Nucl. Acids Res., 17, 6369 (1989).
35. G.D. Stormo, T.D. Schneider, L. Gold, and A. Ehrenfeucht, Nucl. Acids Res., 19, 2997 (1982).
36. M. Kudo, Y. Lida, and M. Shimbo, CABIOS, 3, 319 (1987).
37. K. Nakata, M. Kanehisa, and Ch. DeLisi, Nucl. Acids Res., 13, 5327 (1985).
38. S. Brunak, J. Engelbrecht, and S. Knudsen, J. Mol. Biol. 220, 49 (1991).
39. J. Engelbrecht, S. Knudsen, ang S. Brunak, J. Mol. Biol., 227, 108 (1992).
40. E.C. Uberbacher and R. Mural, Proc. Natl. Acad. Sci., 88, 11261 (1991).
41. R. Farber, A. Lapides, and K. Sirotkin, J. Mol. Biol., 226, 471 (1992).
42. E.E. Snyder and G. Stormo, Nucl. Acids Res., 21, 607 (1993).
43. N. Qian and T.J. Sejnowski, J. Mol. Biol. 202:865-884 (1988).
44. L. Holley and M. Karplus, Proc. Nat. Acad. Sci. USA 86:152-156 (1989).
45. D. G. Kneller, F.E. Cohen, R. Langridge, J. Mol. Bol. 214:171-182 (1990).
46. H. Bohr, J. Bohr, S. Brunak, J. M. R. Cotterill, B. Lautrup, L. Norskov, H. O. Olsen, S. B. Petersen, FEBS Lett. 214:233-228 (1988).
47. B. Rost, C. Sander: J. Mol. Biol. 232, 584-599 (1993).
48. C. Sander and R. Schneider, Proteins 9:56-68 (1991).
49. S. R. Holbrook, S. M. Muskal, S.-H. Kim, Protein Engeneering 3:659-665 (1990).
50. S. M. Muskal, S. R. Holbrook, S.-H. Kim, Protein Engeneering 3:667-672 (1990).
51. J. D. Hirst and M. J. E. Sternberg, Protein Engineering, 4:615-623 (1991).
52. M. O. Dayhoff, R. M. Schwarz, B. C. Orcutt, A model of evulutionary change in proteins, In Atlas of Protein Sequence and Structure (M.O. Dayhoff ed.) vol. 5, suppl. 3, p. 345, National Biomedical Research Foundation, Washington, DC (1978).
53. R. C. Wade, H. Bohr, P. G. Wolynes, J. Am. Chem. Soc. 114:8284-8285 (1992).
54. M. S. Friedrichs, R. A. Goldstein, P. G. Wolynes, J. Mol. Biol. 222:1013-1034 (1991).
55. H. Bohr, J. Bohr, S. Brunak, R.M.J. Cotterill, H. Fredholm, B. Lautrup and S.B. Pertersen, FEBS-Letters 261:43-46 (1990).
56. E. A. Ferrán and P. Ferrara, Biol. Cybern. 65:451-458 (1991).
57. E. A. Ferrán and P. Ferrara, CABIOS 8:39-44 (1992).

STATISTICAL MODELS OF CHROMOSOME EVOLUTION

M.J.Bishop

HGMP Resource Centre, Watfod Road
Harrow, Middlesen HA1 3UJ, UK

J.H.Edwards and J.L.Dicks

Genetics Laboratory, Dept. of Biochemistry, University of Oxford
South Parks Road, Oxford OX1 3QU

1. INTRODUCTION TO PHYLOGENETIC INFERENCE

Populations of biological organisms change with time and we can study how their genetic material is reassorted at meiosis and transmitted from one generation to the next [1]. Over longer time spans we cannot usually observe the changes in the genetic material directly but we can study the genetic properties of extant populations which we assume to have diverged from each other [2]. DNA sequencing is a direct way of characterising the genetic constitution of an individual and DNA can be recovered from bones dated by stratification in archaeological sites. Studies of fossil DNA, for example from Miocene leaves, have been reported. For the majority of extant species we are unlikely to be able to study the DNA of their antecedents. The comparative method in biology has been pursued for several hundred years on morphological criteria and for several decades on molecular criteria. The ideas of Darwinian evolution came out of a consideration of a combination of animal and plant breeding, adaptation of organism to environment, and comparative studies both morphological and biogeographical. The aim of phylogenetic inference is narrower. It attempts to elucidate the order of descent of organisms from common ancestors, most commonly on a tree. A popular example is of the three primates man, chimp and gorilla. There are three possible ways these animals could be related on a tree: (man (chimp, gorilla)); ((man, chimp) gorilla) and ((man, gorilla) chimp). Different datasets and different methods of analysis indicate different answers to the problem (a present consensus of these suggests the relationship is the second listed). If phylogenetic analysis is to be more than idle speculation then a statistical foundation to the subject is needed with explicit assumptions and testable hypotheses. Most progress has been made in this with regard to DNA sequences but other sorts of information are amenable to similar treatment. In fact the work we are doing concerns chromosomes and the orders of genes along them, but it is helpful to review other work first.

2. DATA FOR PHYLOGENETIC INFERENCE

In any field of comparative biology (with or without an evolutionary flavour) one needs to enumerate similarities and differences between individuals, populations and species. It is easy to be tricked by convergent evolution: that is things coming to resemble one another from unrelated lines of descent and this is as much a problem for molecular as for morphological data. The scale of measurement has to be explicit and it must be possible to cover all outcomes. For DNA sequences there can be nucleotide substitution and shrinking or growing by deletion or insertion. It is easy to score the differences between similar sequences in these terms and to enumerate the steps by which one might have changed into another. Size and shape of skulls can be measured and the transformation of one into another described in spatial coordinates. Morphological characters are hard to place in a single scale and one invokes character weighting as an artificial means of doing so in the construction of taxonomies.

Such artifices are meaningless for phylogenetic inference. Molecular data tend not to suffer from this disadvantage of not knowing how to measure like things and much can be related to the underlying DNA sequence (although we do not understand how to do this for protein three dimensional structures). Sequence data are still hard to generate in sufficient quantity in spite of the various genome projects but an advance in sequencing methodology could revolutionise the position.

The karyotype, the genetic linkage map, DNA sequences, protein sequences and protein three dimensional structures can all be expected to contain historical information. Biological macromolecules serve to specify the development and function of the organism not to record historical information. We cannot expect to obtain the answers to all our phylogenetic questions but hopefully many problems will be tractable.

To illustrate the range of possibilities here are a few examples of molecular evidence which has been used to elucidate the relationships of primates.

Cell surface polymorphisms have long been available in the ABO system in man. Chimps have A and O (O is rare in other primates) and gorilla has B only.

The major histocompatibility antigen system (MHC) has perhaps 100 linked loci many of which have multiple (as many as 20) alleles. There is a considerable similarity in the MHC system between man and chimp, less between man and gorilla.

Chromosome staining techniques give banding patterns, man/chimp retaining 13 identical chromosome pairs, man/gorilla 9 and man/orang 8. It is apparent that blocks have been shuffled.

Enzyme polymorphisms can be readily studied by gel electrophoresis. Of 23 enzymes studied in man, chimp and gorilla there are so few differences that phylogenetic information is not forthcoming.

Immunological cross reactions were the first line of evidence that man, chimp and gorilla formed a group with orang being more distant. Previously it had been assumed that man was more distant from the other animals.

DNA hybridisation data (reannealing rates in mixed populations of molecules) support the relationship ((man, chimp) gorilla).

3. MODELS FOR PHYLOGENETIC INFERENCE

A tree model for the relationships of organisms is the first prerequisite in any attempt to infer species phylogenies. The tree has the leaf nodes representing the extant organisms with the internal nodes representing the splitting on the lineages leading to the present species. A vertical axis represents time in the units of the process of change being studied and can be scaled to sidereal time if a splitting point can be estimated from the fossil record. There is a computational problem which results from the fact that there is no way of finding the best tree shape and times to match the data under consideration except by evaluating all trees (this is called an NP-complete problem by computer scientists). Phylogenetic inference is therefore very difficult for more than a few species because of the exponential growth in possibilities (for 10 species there are more than 10 million trees to evaluate). There are ways of avoiding evaluating trees which are sure to be loosers (branch and bound algorithms) and opportunities for parallel processing (10 million processors each working on an instance could evaluate the trees in one unit of time). The tree model is a good model for sexually reproducing organisms without inbreeding (as in mammals) and can be derived by a large scale collapsing of the genealogy. With inbreeding and separation and subsequent joining of populations closed loops are introduced into the genealogical graph and the estimation problem becomes more complex. Migration confounds the study of the evolution of human populations. When a 'molecular clock' stochastically constant process of evolutionary change is not imposed as part of the model an unrooted tree is the appropriate graph.

The other component required is the model of change of the property being studied. The model has to be appropriate to the data and includes the assumptions being made about the evolutionary behaviour of the property. In the case of human population gene frequencies for blood groups Thompson [2] used a Brownian motion approximation for random genetic drift as the statistical model. Felsenstein [3] used a Poisson process model of mutation for the evolution of DNA sequences (in the absence of insertion and deletion). These models are simplification of reality but can be refined as we obtain more understanding of the processes involved. Without a model, no statistical inferences are possible. Inferences are necessarily made within the framework of the model and it is the model which should be the subject of criticism rather than the inferences themselves (conflicting inferences of phylogeny for the same species with different data give a warning signal).

Likelihood inference has been the method of choice for phylogenetic studies [4]. Likelihood theory advocates that the merits of hypotheses are ranked according to their associated likelihood on the same data. The likelihood of an hypothesis H given data D is

$$L(H) = P(D|H).$$

All the information in the data on the relative merits of two hypotheses is contained in the likelihood ratio

$$L(H1) / L(H2).$$

In the phylogenetic inference problem one is looking for the tree shape and times of the internal nodes which give the highest likelihood. This is a problem in maximum likelihood estimation. It is possible that there may be no single clear winner.

It is important to investigate the fit of the data to the model and not to assume that one has chosen a good model first time round. Goldman [5] has devised useful tests of the fit of data to model for the case of DNA sequences.

It should be mentioned that the popular parsimony estimates of phylogeny (which seek to find the tree topology which minimises the amount of evolutionary change) are inferior because the model is implicit not explicit, are not statistical, and do not involve time.

4. PROGRESS IN MODELS FOR ESTIMATION OF PHYLOGENY FROM SEQUENCE DATA

The earliest models for DNA sequence evolution [3] were incomplete because they did not consider insertion and deletion. It should be possible to infer phylogenies and multiple sequence alignments from DNA sequences in a single operation. This goal has not yet been attained. Bishop & Thompson [6] were pioneers in the inclusion of insertion and deletion into the model. Their method was much improved by the work of Thorne et al. [7] who represented the insertion-deletion process as birth-death process of imaginary links that separate single nucleotides. Thorne et al. [8] made a further improvement by considering fragments of DNA being inserted or deleted rather than single nucleotides. These methods work for pairs of sequences and could be extended to joint estimates for multiple sequences. Similar approaches are applicable to the study of chromosomal evolution.

5. CHROMOSOMES AND LINKAGE GROUPS

Chromosomal evolution has added dimensions over the problem of DNA sequence evolution. Some groups of genes are controlled together and need to be physically adjacent. Others can be located anywhere and we see shuffling of blocks of DNA both within and between chromosomes. Telomeres are important structures to prevent loss of DNA from chromosome ends. Centromeres seem less important and can come and go.

Eventually we hope to know the complete sequence of the large chromosomes of many higher organisms. At present we know the complete sequence of the much smaller organelles (mitochondria and chloroplasts). DNA coding for structural RNA or for protein is usually the only parts of the genome which are adequately preserved across distant taxa to permit phylogenetic inference. However, these molecules are under considerable selective pressure and must be able to perform their cellular function. Even a single point mutation in double dose can be lethal (sickle cell anaemia is the classic example).

There are doubtless some constraints on gene order but these are less than the constraints on the sequences of functional macromolecules. Gene order is another line of evidence which can be used for phylogenetic inference [9]. By 'moving up' a level to gene order rather than gene sequence there may be some advantages in that selective effects are reduced and computation is simpler. Gene order is most directly obtained by the tedious task of complete sequencing, as in the case of the mitochondrial genomes, but a variety of other techniques enable gene order to be established to varying degrees of confidence.

Genetic mapping by linkage and haplotype analysis enables the order of loci to be established along the chromosome. Mouse back cross breeding programmes involving the sibling species Mus musculus and M. spretus give DNA which can be unambiguously typed by either RFLPs in Southern blot analysis or VNTRs in PCR analysis. This procedure is very informative about marker order. Comparative studies of maps from different species have shown that there are linkage groups conserved across vertebrates [10]. A review of the conservation of linkage in mammals is given by O'Brian et al. [11]. The most extensively studied pair of species are man and mouse [12]. Interest in mapping genes in a variety of other mammals (pig, sheep, ox, horse, dog) now gives the opportunity for further studies.

Cytological properties of chromosomes have long been studied in relation to animal evolution [13]. The number and morphology of chromosomes varies considerably from species to species. Bands on the giant chromosomes from dipteran salivary glands were noted by Balbiani in 1881 (though he did not realise the structures are chromosomes). Rearrangements of these banding patterns in sibling species of Drosophila have enabled the phylogenetic relationships to be inferred [14]. Banding patterns in the chromosomes of mammals are revealed by the use of dyes [15]. This method makes possible the identification of each of the 23 human chromosomes and the detection of chromosomal abnormalities. The phylogenetic relationships of the primates have been inferred from banding patterns of the chromosomes [16]. Visible changes in chromosomes arise frequently and are called chromosome aberrations [17]. The changes can involve chromosome parts, whole chromosomes or whole chromosome sets. Changes in chromosome structure are classified as

deletions,
duplications,
inversions,
translocations.

Any aberration is a group of chromosome segments joined together according to certain basic rules. When aberrations become fixed in a populations they lead to the evolution of a new stable karyotype. The evolutionary model describes the transformation of one karyotype into another by the same basic rules applied over time. The position of centromeres is also subject to change.

The work we are doing involves the development of statistical models for the processes of chromosome evolution, and of tests to check whether a particular combination of tree and chromosomal evolution model is an acceptable explanation of the observed chromosome segment data.

The data are the arrangements of the segments in the extant species. Segments may be characterised by their banding patterns (Drosophila, primates). The models are being used to infer phylogeny for these data. However, the recently available data on the order of molecular markers is a much more powerful method of characterising segments and the main part of the study involves these.

There is an interesting complication in the use of molecular markers which to be adequately conserved over the species of interest will normally be genes coding for proteins. The sequences themselves are evolving and will not be identical in the different species. Genes are also duplicating and deleting to form multigene families of considerable complexity. Visible chromosomal duplications are not the only cause of the proliferation of multigene families. Other processes may be involved such as tandem duplication of individual genes, gene conversion (alteration of part of one copy of a gene by another member of the family), and reverse transcription (incorporation of RNA messages back into genomic DNA).

For single copy genes there is no problem. However, in the case of multigene families a careful study of the members in the species of interest is a pre-requisite for accurate labelling of similar segments from different karyotypes. Intraspecific studies of the distribution of members of multigene families are meaningful. [The two concepts may be described as 'species phylogeny' of 'orthologous' genes between species and 'gene phylogeny' of 'paralogous' genes within species.] To complicate matters further there are pseudogenes which are supposed to be non-functional relics of members of gene families. And finally, some genes are composed of segments of numbers of other genes.

The problem of the identification of conserved segments in chromosomes labelled by unique markers is easier than the problems of aligning molecular sequences, written language, or bird song [18]. If the markers cannot be uniquely labelled (doubt about gene family relationships) the problem becomes harder but is still soluble. [We may expect relationships within multi-gene families to be clarified in some cases.]

We can compare the genomes of two species by means of an Oxford Grid [19]. Each row or column represents a chromosome, the larger the chromosome, the wider the row or column. Each homologous locus is plotted in the relevant cell, each cell being referenced by its chromosome number in each of the two species. To give a general impression of the similarity of two species, it is simplest to plot each locus as a point in the cell. A cell with a high point density may denote a large conserved segment of DNA between the species. For examples of this type of grid, see the Human-Chimpanzee (Fig. 1), Human-Cow (Fig. 2) and Human-Mouse (Fig. 3) grids. It is immediately clear that the number of chromosomal changes in the evolution of these species pairs differs enormously. We would like to model this evolution and be able to estimate the evolutionary times involved.

For a more informative grid, when looking separately at each cell, having `zoomed in', we can draw diagonal lines, rather than individual points, the cells scaled so that the length of the line will directly show the length of the conserved chromosome segment. The orientation of the diagonal indicates if the segment is inverted or not.

Another useful form of display is to choose a chromosome of a first species and place it along the vertical side of a rectangle. The corresponding fragments of chromosomes of the other species are displayed along the horizontal and diagonals drawn as before. A complete

Human

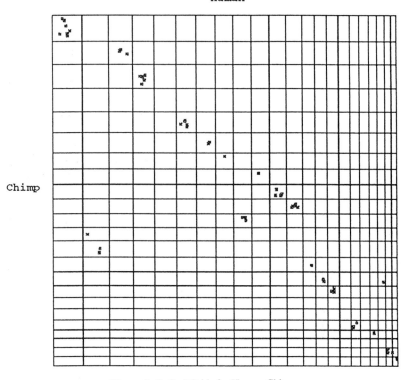

Figure 1. Oxford Grids for Human-Chimp

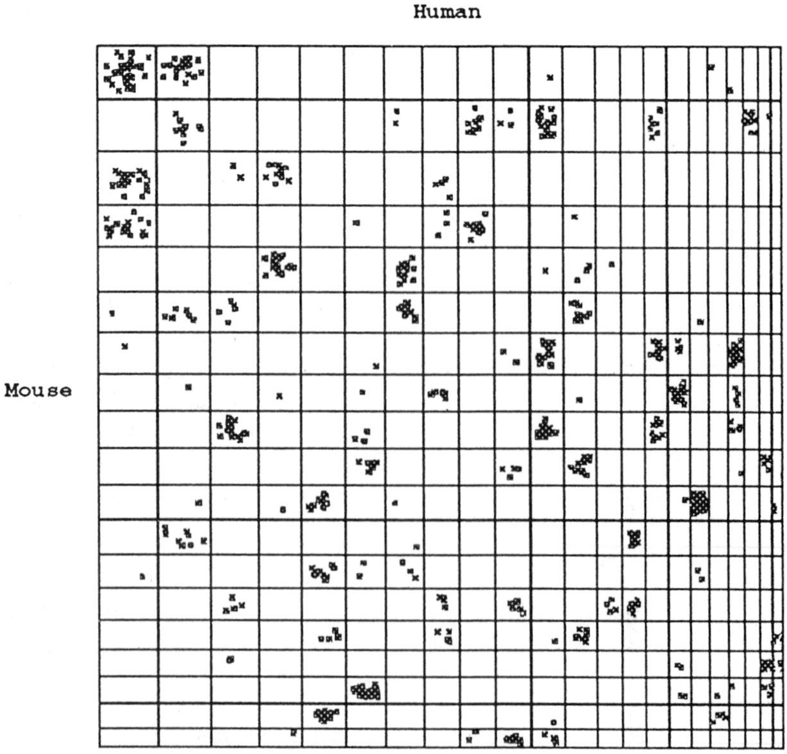

Human

Mouse

Figure 2. Oxford Grids for Human-Mouse

214

Human

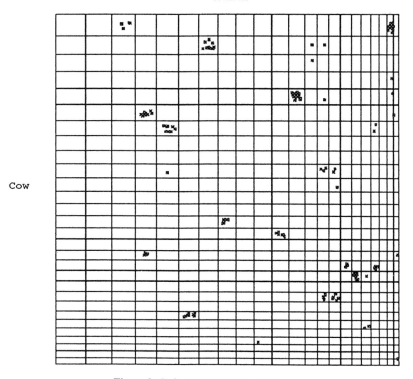

Figure 3. Oxford Grids for Human-Cow

diagonal should be produced when adequate data are available (orientation of the fragments can be flipped if necessary). The gene names should be included on the diagram and one would want to choose regions to zoom in and out. This is comparable to an alignment of sequences.

6. PHYLOGENETIC INFERENCE FROM CHROMOSOME DATA

An important method is to calculate the minimum number of reciprocal translocations and inversions (whether pericentric or paracentric) required to convert one genome to another (simplest case - currently ignoring duplications, deletions and the very rare transpositions). This has already been done for a special case of the grid by Sankoff et. al [9]. They looked at the gene order for mitochondrial genomes. This problem corresponds to an Oxford Grid with one cell. The extended case is obviously more complex although similar methods will be used.

The method of inference being used is maximum likelihood. Refined segment data will be produced after examination of the information with the display tools. Tree patterns and extent of divergence will be estimated from the segment data under the evolutionary models. The movement of chromosomal segments is framed in terms of a birth-death process of the links [8]. Validity of the models will be checked by Monte Carlo methods.

REFERENCES

1. Cannings,C., & Thompson,E.A. 1981. Genealogical and genetic structure.Cambridge University Press: Cambridge.
2. Thompson,E.A. 1975. Human evolutionary trees. Cambridge University Press: Cambridge.
3. Felsenstein 1981. J. Mol. Evol. 17, 368.

4. Edwards A.W.F 1972. Likelyhood, Camebridge Univ. Press.
5. Goldman,N. 1991. Statistical Estimation of Evolutionary Trees. Ph.D. thesis, University of Cambridge.
6. Bishop,M.J., Friday,A.E. & Thompson,E.A. 1987. In Bishop,M.J & Rawlings,C.J.eds. Nucleic acid and protein sequence analysis. IRL Press: Oxford.
7. Thorne,J.L., Kishino,H. & Felsenstein,J. 1991. J. Mol. Evol. 33, 114-124.
8. Thorne,J.L., Kishino,H. & Felsenstein,J. 1992. J.Mol. Evol. 34, 3-16.
9. Sankoff,D., Cedergren,R. & Abel,Y. 1990. Methods in Enzymology 183, 428-438.
10. O'Brian,S.J. ed. 1987. Genetics maps. CSH Laboratory: New York.
11. O'Brian,S.J., Seuanez,H.N. & Womack,J.E 1988. Ann. Rev. Genet. 22, 232-251.
12. Searle,A.G. et al. 1989. Ann. Hum. Genet. 53, 89-140.
13. White,M.J.D. 1973. Animal cytology and evolution. Cambridge University Press: Cambridge.
14. Carson & Kaneshiro 1976. Ann. Rev. Ecol. Systematics 7, 311.
15. Bickmore,W.A. & Sumner,A.T. 1989. TIG 5, 144-148.
16. Yunis,J.J. & Prakash,O. 1982. Science 215, 1525-1529.
17. Suzuki,D.T., Griffiths,A.J.F. & Lewontin,R.C. 1981. An introduction to genetic analysis. W.H.Freeman: San Francisco.
18. Sankoff,D. & Kruskal,J.B. 1983. Time warps, string edits and macromolecules.Addison-Wesley: Reading, Mass.
19. Edwards,J.H. 1991. Ann. Hum. Genet. 55, 17-31.

DECIPHERING THE GENETIC MESSAGE: SOME THOUGHTS ON PRESENT LIMITS

C. Frontali

Laboratory for Cell Biology
Istituto Superiore di Sanità
Rome, Italy

The analogy with a written language is often used with reference to the genetic message contained in DNA, and is reflected in the use of terms like transcription, translation etc. In some studies the concepts of grammar and syntaxis have also been introduced. The analogy is extended further in statements claiming the possibility of identifying DNA with the Rosetta stone.

Analogies have an important role in scientific thinking and modelling, but they might be severely misleading if one is not aware of their limits. Here I want to develop some considerations on the acritical use of a tacitly assumed analogy with human languages in genome research. The aim of these considerations is to identify areas where more elaboration is needed in the important task to correctly store and analyse the gathered information.

In modelling our understanding of the genetic message after the structure of a text written in one of our languages, we implicitly make a strong hypothesis concerning the degree of linearity of the message. In effect we often imagine the genetic "text" as a written page (or a book), where a series of "words" is linearly arranged. Algorithms have been developed to identify meaningful words. In doing so, it is also (often unexplicitly) assumed that genetic words are in a 1:1 correspondence with their meaning, so that an hypothetical vocabulary would allow unambiguous decodification of any part of the text.

This situation is actually present in coding regions. Here the set of t-RNAs represents the vocabulary indicating the correspondence between message units (codons) and product building blocks (amino-acids). The relatively early discovery of a universal genetic code (to which only very few exceptions have been found) most probably influenced our thinking, and, being in line with our usual way to codify information, was often acritically extended to those large portions (more than 90% of the genome) which are not translated into proteins.

Let us first stress an important point, still using the same linguistic analogy: co-linearity between protein and nucleic acid sequences in coding regions provided us with the bilingual inscription (the Rosetta stone) necessary for deciphering the system of codons. In non-coding regions we are confronted with a completely unknown language, and we lack a bilingual inscription relating it to its functional meaning.

Paradoxically, if nucleic acid sequencing had been a cheaper and easier task, and complete genome information at the nucleotide level had been stored in computers before a number of biological functions (e.g. promoters) had been identified, it would be now extremely difficult to retrieve and utilise this information. The idea to accumulate a constantly increasing number of sequence elements without any indication as to their presumed function is a rather naive approach to the decodification problem: no language was ever deciphered only through the study of inter-symbolic correlation (not even Cretan linear B system, which was luckily identified with a known language).

This leads us to consider that functional annotation is an essential feature of a useful database, providing keys for the retrieval of sequences which belong to the same functional class (e.g. promoters, silencers, enhancers, origins of replication etc.) but do not necessarily possess a significant degree of similarity in nucleotide sequence.

Functional data (the second, known language of the desired bilingual inscription) should thus ideally form an essential part of each database entry: The problem is that we do not presently know all the possible functions of non-coding DNA, and while new sequences are constantly added to the database, new functional information on already deposited ones is not, which makes that our sequence databases are progressively loosing value. A truly integrated database should not only connect information obtained at different levels of topographical resolution, but also foresee continual up-dating of the annotation in terms of functions newly identified through genetic or molecular data. Research is needed to devise the best way to achieve a dynamic annotation system, susceptible of up-dating with increasing knowledge on the content, type, and number of essential annotation items.

Another critical point concerns the assumption of linearity versus context-dependence of the information contained in non-coding portions. Are we sure that defined groups of nucleotides form the meaningful units of the unknown language? One might also imagine that deviations from standard DNA configuration (possibly realised by different local sequences) are the real characters in which the message is written. The reading mechanism might well be a rather loose one, in which a given DNA element is recognised by different factors (or combinations of factors) and assumes different meanings according to its environment or context. Further, non-linear effects (absent in a book) are originated by DNA flexibility in space, allowing interactions between sequence elements widely separated along the molecule.

Among future developments in biocomputing methods for genome analysis, we need to devise tools to deal with such problems, since additivity and linearity of the genetic message in non-coding regions cannot be given for granted.

CONTRIBUTORS

Jaques Beckmann
  Centre d'Etude du Polymorphisme Humain, 27 rue Joliette Dodu, 75010 Paris, France

Martin J. Bishop
  University of Cambridge, Computer Laboratory, Corn Exchange Street, Cambridge CB2 3QG, UK

Elbert Branscomb
  Human Genome Center, Biomedical Sciences Division, L-452, LawrenceLivermore National Laboratory, 7000 East Avenue, Livermore: CA 94550, USA

Charles Cantor
  Center for Advanced Biotechnology, 6 Cummington Street, Boston, MA 02215, USA

Roy Cantu III
  Human Genome Center, Biomedical Sciences Division, L-452, LawrenceLivermore National Laboratory, 7000 East Avenue, Livermore: CA 94550, USA

Richard M. Durbin
  MRC Laboratory of Molecular Biology, Hills Road, Cambridge CB2 2QH, UK

Clara Frontali
  Laboratory for Cell Biology, Istituto Superiore di Sanitá, Rome, Italy

Gaston Gonnet
  ETH Zürich, Institut für Wissenschaftliches Rechnen, 8092 Zürich, Switzerland

Frank Herrmann
  German Cancer Research Center, Im Neuenheimer Feld 280, 69120 Heidelberg, Germany

Hans Lehrach
  Imperial Cancer Research Fund, Lincoln's Inn Fields, London WC2A 3PX, UK

Jean Thierry-Mieg
  CNRS Physique Mathematique and CRBM, BP 5051, 34033 Montpellier, France

Richard Mott
  Imperial Cancer Research Fund, Lincoln's Inn Fields, London WC2A 3PX, UK

Victor Markowitz
  Lawrence Berkeley Laboratory , 1 Cyclotron Road, Berkeley: CA 94720, USA

219

Eugene Myers

University of Arizona, Department of Computer Science, Gould Simpson Building, Tucson: Arizona 85721, USA

William R. Pearson

University of Virginia, Department of Biochemistry, Charlottesville , Virginia 22908, USA

Martin Reczko

German Cancer Research Center, Im Neuenheimer Feld 280, 69120 Heidelberg, Germany

Jens G. Reich

Max-Delbrück-Centrum für Molekulare Medizin, Robert-Rössle-Straße 10, 13125 Berlin-Buch, Germany

Otto Ritter

German Cancer Research Center, Im Neuenheimer Feld 280, 69120 Heidelberg, Germany

Robert J. Robbins

Johns Hopkins University, Welch Laboratory for Applied Research, 1830 East Monument Street, Baltimore MD 21205, USA

Tom Slezak

Human Genome Center, Biomedical Sciences Division, L-452, LawrenceLivermore National Laboratory, 7000 East Avenue, Livermore: CA 94550, USA

Sándor Suhai

German Cancer Research Center, Im Neuenheimer Feld 280, 69120 Heidelberg, Germany

William R. Taylor

Laboratory of Mathematical Biology, National Institution for Medical Research, The Ridgeway,Mill Hill, London NW7 1AA, UK

Mark Wagner

Human Genome Center, Biomedical Sciences Division, L-452, Lawrence Livermore National Laboratory, 7000 East Avenue, Livermore: CA 94550, USA

Mimi Yeh

Human Genome Center, Biomedical Sciences Division, L-452, LawrenceLivermore National Laboratory, 7000 East Avenue, Livermore: CA 94550, USA

Günther Zehetner

Imperial Cancer Research Fund, Lincoln's Inn Fields, London WC2A 3PX, UK

INDEX

A

B

C